Combinatorics II Problems and Solutions: Counting Patterns

Stefan Hollos and J. Richard Hollos

Combinatorics II Problems and Solutions: Counting Patterns
by Stefan Hollos and J. Richard Hollos

Abrazol Publishing
an imprint of Exstrom Laboratories LLC
662 Nelson Park Drive, Longmont, CO 80503-7674 U.S.A.

Publisher's Cataloging in Publication Data
Hollos, Stefan
Combinatorics II Problems and Solutions: Counting Patterns / by Stefan Hollos and J. Richard Hollos
p. cm.
Includes bibliographical references and index.
Paper ISBN: 978-1-887187-32-9
Ebook ISBN: 978-1-887187-33-6
Library of Congress Control Number: 2016916311

1. Combinatorial enumeration problems. 2. Combinatorial analysis.
I. Title. II. Hollos, Stefan.
QA164.8
511.62 HOL

Contents

PREFACE

This is a continuation of our combinatorics problem book. It deals mainly with pattern counting problems. These are more advanced problems, in that the mathematics behind them is probably a bit more abstract and not as intuitively easy to grasp, at least not with an initial encounter. This does not mean that they are technically more difficult, but they do require that ill-defined and somewhat nebulous concept of mathematical sophistication. This really just means being comfortable with applying mathematical machinery purely in the abstract, i.e. having trust in the machinery.

We give only a brief introduction to the mathematics behind these problems. The emphasis is on using the mathematics and not on its theoretical foundation. In our experience, when you discover the power of some new mathematical technique, and become adept at using it, then the motivation for understanding its foundation arises naturally. It is more difficult to find this motivation if you don't really understand what the math can do for you. At least that is the case with us, and it may be the result of a natural bias as physicists, who use mathematics from a more utilitarian point of view. There is a list of references for readers who want to dig deeper into the theoretical foundations of these techniques.

Here is a simple example of the kind of pattern counting we are interested in. We have black and white square tiles and we want to create a linear arrangement of n of them. With no restrictions, elementary combinatorics tells us there are 2^n possibilities. Now if we do not allow two or more black tiles to appear side by side, how many arrangements are possible? The beginning combinatorialist may have a hard time seeing how to tackle a problem like this. Several ways to solve problems of this kind are covered in the book.

The patterns we talk about are usually represented as a string of symbols. A class of patterns is represented by a set of strings. The set of strings is sometimes called a language. Many pattern classes fall into the category of regular languages. These are languages that can be generated by a finite automaton. A regular expression is a shorthand way of describing a regular language. Regular expressions and finite automata are very powerful ways to describe, generate, and count patterns. This book describes these tools and shows how to use them.

Another type of pattern counting problem involves equivalence under symmetry. Suppose we want to create a necklace using beads of three colors. An elementary analysis says there 3^n ways to make an n bead necklace. But many of these necklaces will be related by a simple circular shift of the beads. If such necklaces are considered equivalent and counted as one, then how many unique necklaces are there? What if we also equate necklaces where one is just another that is picked up and turned over? These problems are surprisingly easy to answer using a method called Polya's theory of counting. We cover this method and its more general form, called Burnside's theorem. There are many worked out problems that show how to use the method. Included are problems that find the number of unique ways to color the Platonic solids.

In short, this book describes, and shows how to use, very powerful methods for counting patterns. These are methods that should be in the toolbox of every combinatorialist. It not only shows how to count them but also provides the means to generate them with programs that can be downloaded from the book's web page on the Abrazol website at
abrazol.com

May you have many happy hours counting and generating patterns.

Stefan Hollos and Richard Hollos
Longmont, Colorado

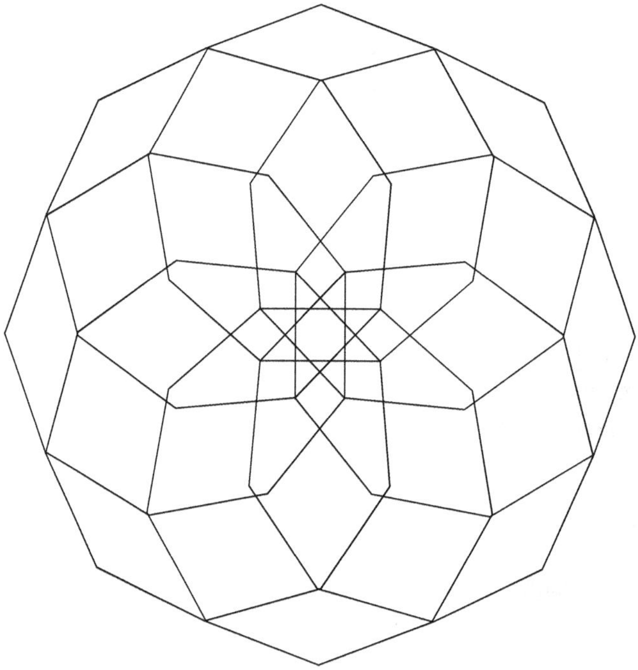

COUNTING WORDS

We define a word as a sequence of n letters that are chosen from an alphabet with an infinite supply of m different types of letters. The term sequence means that there is an order to the letters, similar to the way events are ordered in time. There is a first letter, a second, and so forth. A letter is distinguished by its type and position in the word. Another way of saying this is that we have n distinguishable boxes and we want to put one of m different types of objects into each box.

The simplest question is: How many different words can we create? There are m possibilities for each of the n letters so taking the product we get m^n different words. But what if we fix the number of letters of a given type? How many words are possible then?

This is still a fairly simple question but let's make it even simpler by looking at the case where $m = 2$ and the letters are x and y. How many words are possible when k of the letters are y? This is equivalent to asking for how many ways k objects can be chosen from a set of n objects, where the order in which they are chosen does not matter. This is one of the most basic questions in combinatorics and the answer is of course $\binom{n}{k}$.

The reason for looking at these simple questions is that it leads to a nice symbolic solution method that can also be used for more difficult problems. For example, we can represent the last question symbolically as a product of n binomials.

$$(x+y)(x+y)\cdots(x+y) = (x+y)^n \tag{1}$$

If the x and y symbols, representing the letters in the alphabet, are treated as regular algebraic variables and the expression is multiplied out, then the coefficient of $x^{n-k}y^k$ will be $\binom{n}{k}$, the number of words with k y's and $n-k$ x's. The expression $(x+y)^n$ is called a generating function for n letter words composed of the letters x' and y. Note that it also gives the total number of different words. If we set $x=y=1$ then $(x+y)^n = 2^n$. Also, if we are interested only in the number of x's in a word then set $y=1$. The coefficient of x^k in $(x+1)^n$ is the number of words containing k x's.

Instead of interpreting equation 1 as a common algebraic expression we can interpret it as something called a regular expression. This means interpreting addition as a logical OR operation and multiplication as a logical AND operation. As a logical expression, the equation says the first letter can be x or y and the second letter can be x or y and so on. This simple change in perspective is very powerful and can be used to create a versatile pattern matching formalism which we will discuss in the next section.

REGULAR EXPRESSIONS

A regular expression is used to define a set of strings constructed from a finite alphabet of letters or symbols. The set of strings so defined is called a regular language. The best way to understand what this means is by example. Let the alphabet consist of four letters $A = \{a, b, c, d\}$. One simple language is all the two letter strings that begin with the letter a. The regular expression defining the language is: $R = aa+ab+ac+ad$. The $+$ operator in the expression is analogous to a logic *or* operator or a set theory union operator. The expression can be simplified a bit by writing it as $R = a(a + b + c + d)$ which represents the concatenation of a with any of the possible one letter strings. Concatenation is analogous to a logic *and* operator.

This example shows the two basic operations used to construct regular expressions: union and concatenation. If R and S are regular expressions then $R+S$ and RS are also regular expressions. The simplest regular expression is just a single letter, $R = a$ or $R = \epsilon$ where ϵ is an empty string i.e. a string with no letters. Multiple concatenations can be written in a manner analogous to exponentiation, $ababab = (ab)^3$ or if $R = ab$ then $RRR = R^3$. Note that $(ab)^3 \neq a^3b^3 = aaabbb$ since concatenation is not commutative.

Another operator used in regular expressions is called

the Kleene star. It is defined as follows: $R^* = \epsilon + R + R^2 + R^3 + \cdots$. In other words, a starred regular expression can appear zero or more times concatenated with itself. Using the Kleene star, the regular expression for all possible strings in the alphabet $A = \{a, b, c, d\}$ is simply $R = (a + b + c + d)^*$. This says the first letter of the string can be any one of a, b, c, or d, the second letter can be any of those same letters and so on for any length string. As another example consider the regular expression $R = (0 + 10)^*$. This describes the set of binary strings in which a 1 never appears two or more times consecutively.

For our purposes this sums up how regular expressions are defined and what they mean. There is some additional shorthand notation that is sometimes used to simplify regular expressions. For example, the regular expression $0+1+2+3+4+5+6+7+8+9$ can be abbreviated as $[0-9]$ and the letters a through z can be abbreviated as $[a-z]$. Another common abbreviation is $\epsilon + a = a?$. A simple identity that is sometimes useful is $\epsilon + aa^* = a^*$. For more identities, theorems, and other goodies involving regular expressions see one of the references.

The set of strings defined by a regular expression is called a regular language. The regular expression for a language need not be unique. It is possible to have two different regular expressions that describe the same language. You will see why in the next section on finite automata, to which regular expressions are closely tied. There is an algebra for regular expressions that can, in principle, be used to simplify them but it's generally not easy to use and not worth the trouble.

The set of regular languages is infinitely large but there are languages (patterns) that cannot be described by a regular expression. A context free grammar is a more general way of describing a language. Regular languages are a subset of the languages described by context free grammars. In some cases a context free grammar can be expressed as a recursive regular expression or a set of such expressions. An example of a recursive regular expression, for strings over an alphabet $A = \{a, b\}$, is $R = (aRb)^*$. The regular expression is defined in terms of itself, which essentially allows it to be infinitely long. A recursive regular expression describes an automaton with an infinite number of states and it allows the generating function for such an automaton to be easily found. Automata and generating functions are discussed in the following sections.

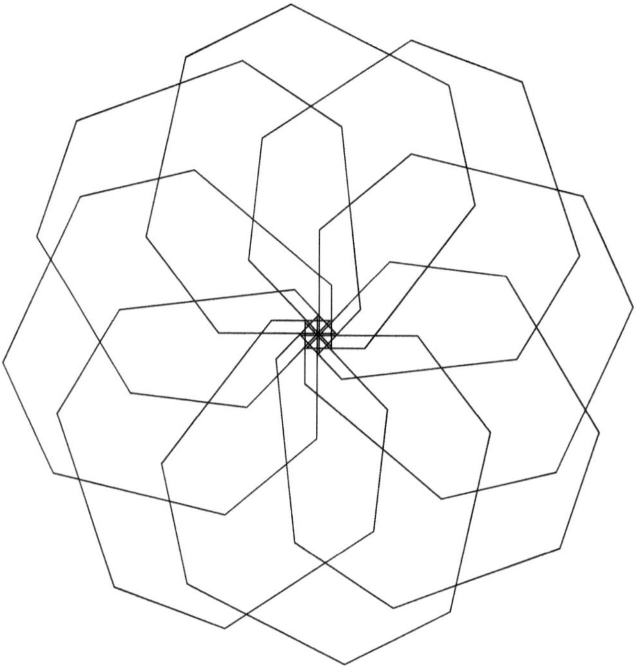

FINITE AUTOMATA

Why are automata important? One reason is that when you need a regular expression, sometimes it's easier to create the automaton, then get the regular expression from that. So being able to make an automaton for what you want is a good skill to have. Secondly, some programs that take regular expressions as input, like grep and flex, will convert that regular expression into an automaton, then use that with a very fast algorithm to perform the search. So it's good to know how regular expressions and automata are related. Thirdly, automata are an elementary computational model and are very important in the theory of computation.

An automaton is a very simple little computer. It has a set of states and it transitions from one state to another according to the sequence of inputs it reads. The inputs are usually represented as a string of symbols or characters. One of the states is the start state and one or more of the states are end states. The start state may also be an end state. The computation it performs is a yes or no answer to the question: does the input sequence move the automaton from the start state to one of the end states? The answer will of course depend on how the states are connected. To see clearly how all this works requires some formal notation and definitions.

Let Q represent the set of states of an automaton and let $q_i \in Q$ be an element of Q. At the end of the book we will introduce some problems that require automata with a countably infinite number of states but most of the automata in this book will have a finite number of states. Let $q_0 \in Q$ be the start state. The automaton always starts in this state. Let $E \subseteq Q$ be the set of end states. If the automaton is in a state $q_i \in E$ at the end of the input then the input is said to be accepted. The end states are also sometimes called accept states or final states. Let Σ be a finite set of symbols or characters called the alphabet. The input sequence will consist of a string of characters taken from Σ. Let $\delta(q_i, a) = q_j$ be the automaton transition function. The function says that if the automaton is in state q_i and the next input symbol is $a \in \Sigma$ then the automaton will move to state q_j. If the automaton is in state q_i and it reads a symbol for which no transition is defined then it stops and the input is not accepted.

Human beings like pictures and for that reason automata are usually represented as directed graphs. Flip through the following pages and you will see lots of pictures of automata. The nodes in the graph represent states and the directed edges represent transitions from one state to another. Each edge is labeled with an alphabet symbol. If $\delta(q_i, a) = q_j$ then there will be an arrow from node q_i to node q_j labeled a. You will sometimes see start nodes designated by a double circle. In this book we will simply state what the start and end nodes are in those cases where it is not obvious.

You will also sometimes see edges labeled with the symbol ϵ. This is the symbol for an empty string and it means that the transition is possible with no input. If an automaton has ϵ transitions or if it has a state with two different transitions for the same input symbol then it is called nondeterministic. Otherwise it is called deterministic. The abbreviations NFA and DFA are used for nondeterministic and deterministic finite automaton. Every NFA can be turned into an equivalent DFA with a larger number of states. For the problems in this book the distinction between an NFA and a DFA will not be important. More information about their differences and how they are related can be found in references at the end of the book.

One of the things we will want to do in the following problems is to characterize the family of strings accepted by an automaton. One way to characterize them is by a regular expression (discussed in the previous section). In some cases, for a small automaton, it is possible to write down the regular expression by inspection, but in general the automaton will need to be simplified by removing some of its states to create a smaller equivalent automaton. This is best explained with an example. Look at the three state automaton in the figure below.

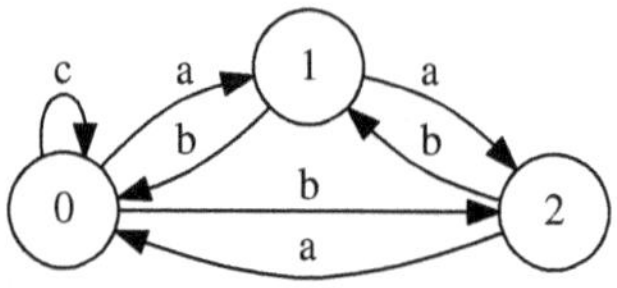

What we want to do is remove state 0, leaving only states 1 and 2. The transitions in the reduced automaton should be such that state 0 is still effectively present. The reduced automaton is shown in the figure below.

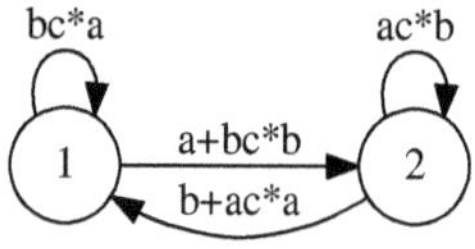

You can see that the transitions are no longer labeled by single symbols. The transition from state 1 to 2 is now labeled $a + bc^*b$ where the $+$ operator is an *OR* operator and the $*$ superscript on the c indicates any number of c's, $c^* = \epsilon + c + cc + ccc + \cdots$. The label means the transition can occur on a single a or on a b followed by any number of c's, followed by another b. It is the bc^*b part of the transition that takes into account the effect of the removed 0 state. You can see from the original automaton that it is possible to go from state 1 to state 2 by first going to state 0 on a b input, staying at state 0 for any number of c inputs and then going to state 2 on another b input. When state 0 is removed this possibility has to be added to the transition from state 1 to 2. In the original automaton you can return to state 1 by going through state 0 on an input of bc^*a so this is taken into account by the loop back transition on state 1 in the reduced automaton.

You could continue in this example by eliminating state 2. This would leave only state 1 with a loop back transition labeled $bc^*a+(a+bc^*b)(ac^*b)^*(b+ac^*a)$. If state 1 is both the start and end state for this automaton then the strings it accepts are the ones that can be generated by going around this loop transition any number of times. The strings are then characterized by the expression $(bc^*a + (a + bc^*b)(ac^*b)^*(b + ac^*a))^*$ which is called the regular expression for the automaton.

Keep in mind that while the language accepted by any finite automaton can be described by a regular expression, the regular expression need not be unique. It is possible to have two different regular expressions that describe the same language. When eliminating states in an automaton, the order in which they are eliminated can sometimes lead to different regular expressions that are equally valid. This is analogous to the fact that you can have two polynomials that are equivalent, with one being a simplified form of another where redundant terms have been canceled.

GENERATING FUNCTIONS

The language accepted by an automaton can have strings of many different lengths. It is useful to know how many strings of a given length are in a language. You can do this by constructing what is called a generating function for the language. We will usually use the notation $G(z)$ for the generating function. For a regular language, i.e. a set of strings accepted by a finite automaton, $G(z)$ will always end up being either a polynomial or a rational function of two polynomials which can be expanded as a power series. Some problems will involve automata with an infinite number of states in which case the language is no longer regular and the generating function may involve square roots of polynomials. The regular expression in this case can sometimes be described recursively. See the section on regular expressions for an example. In any case the coefficient of z^n in the power series expansion of $G(z)$ will equal the number of strings in the language that have length n

$G(z)$ can be derived from the adjacency matrix of an automaton or from the automaton's regular expression. First let's look at how to get it from the regular expression. You can get $G(z)$ from the regular expression by replacing each alphabet letter by z and replacing each starred expression, R^* by $(1 - R(z))^{-1}$ where $R(z)$ is gotten by replacing each alphabet letter in R by z. The

+ operators are then treated as addition operators and concatenation is treated as multiplication.

As an example, we will find the generating function for the number of binary strings that end in 0 and where two or more consecutive 1's never appear. The regular expression that defines these strings is $R = (0 + 10)^*$. 0 is replaced with z and 10 is replaced by z^2. These two terms are in a starred expression, so we take the reciprocal of 1 minus their sum to get the generating function

$$G(z) = \frac{1}{1 - z - z^2} \tag{2}$$

If you expand this as a power series, you get

$$G(z) = 1+z+2z^2+3z^3+5z^4+8z^5+13z^6+21z^7+34z^8+\cdots \tag{3}$$

The power series coefficients are called the Fibonacci numbers. The coefficient 13, of z^6, means that there are 13 binary strings of length 6 that end in 0 and do not have two or more consecutive 1's.

Now let's do the same thing using an adjacency matrix. The automaton shown below generates binary strings where two or more consecutive 1's never appear. The start state and end state are both 0. All the walks of length n on this automaton, that start in state 0 and end in state 0, will generate all the binary strings where two or more consecutive 1's do not occur.

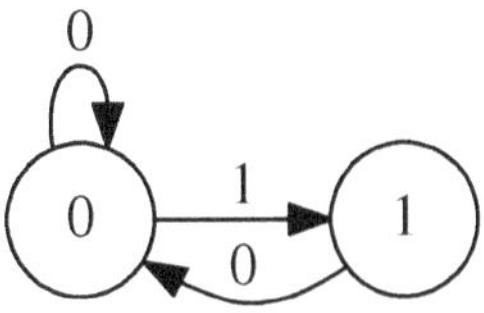

The adjacency matrix[1] for the automaton is

$$\mathbf{A} = \begin{pmatrix} 1 & 1 \\ 1 & 0 \end{pmatrix} \tag{4}$$

Each row and column of an adjacency matrix corresponds to a state. For each transition from state i to state j, we add 1 to element $A_{i,j}$ of the matrix. For this example, state 0 will transition back to itself only on a 1, so we set $A_{0,0} = 1$. State 0 will transition to state 1 only on a 1, so we set $A_{0,1} = 1$. State 1 transitions to state 0 only on a 0 and it never transitions directly back to itself, so we set $A_{1,0} = 1$ and $A_{1,1} = 0$.

The number of walks of length n between state i and state j in an automaton is equal to $(\mathbf{A}^n)_{i,j}$, i.e. the (i, j) element of the nth power of the adjacency matrix. The nth power of the adjacency matrix is equal to the coefficient of z^n in the power series expansion of $(\mathbf{I} -$

[1]This is also sometimes called a transfer or transition matrix.

$z\mathbf{A})^{-1}$

$$(\mathbf{I} - z\mathbf{A})^{-1} = \mathbf{I} + \mathbf{A}z + \mathbf{A}^2z^2 + \mathbf{A}^3z^3 + \ldots \tag{5}$$

This means that for our particular example, the generating function will equal element $(0, 0)$ of $(\mathbf{I} - z\mathbf{A})^{-1}$.

$$(\mathbf{I} - z\mathbf{A})^{-1} = \frac{1}{1 - z - z^2} \begin{pmatrix} 1 & z \\ z & 1 - z \end{pmatrix} \tag{6}$$

The $(0, 0)$ element is $1/(1 - z - z^2)$ which gives us the same generating function that we calculated above.

Now suppose we want not just the number of strings of length n, but the number of strings with k 1's and $n - k$ 0's. To get the generating function for this with the regular expression, we do basically the same thing as before, but instead of replacing both 0 and 1 with z, we replace 0 with x and 1 with y. The term inside the parentheses of the regular expression $R = (0 + 10)^*$ translates to $x + xy$ and since it is starred, we subtract it from 1 and take the inverse to get the generating function.

$$G(z) = \frac{1}{1 - x - xy} \tag{7}$$

Expanding this in a power series we get

$$\begin{aligned} G(z) &= 1 + x + (x^2 + yx) + (x^3 + 2yx^2) + \\ &\quad (x^4 + 3yx^3 + y^2x^2) + (x^5 + 4yx^4 + 3y^2x^3) + \cdots \end{aligned} \tag{8}$$

The coefficient of $y^k x^{n-k}$ in the power series expansion is the number of strings with k 1's and $n-k$ 0's. For example, the number of strings with two 1's and three 0's is the coefficient of y^2x^3, which, from the power series, is 3.

The same thing can be done using the adjacency matrix. If the transition between state i and j is caused by a 0, then set $A_{i,j} = x$, if it is caused by a 1, then set $A_{i,j} = y$. The adjacency matrix then becomes

$$\mathbf{A} = \begin{pmatrix} x & y \\ x & 0 \end{pmatrix} \tag{9}$$

and the generating function is found as before by calculating

$$(\mathbf{I} - \mathbf{A})^{-1} = \frac{1}{1-x-xy} \begin{pmatrix} 1 & y \\ x & 1-x \end{pmatrix} \tag{10}$$

The $(0,0)$ element is $1/(1-x-xy)$ which gives us the same generating function that we got from the regular expression.

The generating function for a recursive regular expression can be found in basically the same way as for a simple regular expression. In the section on regular expressions we gave $R = (aRb)^*$ as an example of a recursive regular expression. If $G(z)$ is the generating function for R then $R = (aRb)^*$ will translate into

$$G(z) = \frac{1}{1 - z^2 G(z)} \tag{11}$$

Solving this equation for $G(z)$ gives:

$$G(z) = \frac{1-\sqrt{1-4z^2}}{2z^2} \tag{12}$$

Expanding this in a power series, you get

$$G(z) = 1+z^2+2z^4+5z^6+14z^8+42z^{10}+132z^{12}+429z^{14}+\cdots \tag{13}$$

The coefficients in the series are called the Catalan numbers. Note that only even powers of z occur in the series. This means all strings in the language must have even length. One way to interpret the language is as the number of ways to create a string of correctly nested parentheses. The letter a corresponds to an open parenthesis and b corresponds to a closed parenthesis. There must always be the same number of open and closed parentheses so only strings of even length are possible. For $n = 6$ the five possible parentheses strings are: ((())), (()()), (())(), ()(()), ()()().

COUNTING NECKLACES

A necklace is a word where the letters are placed in a circular arrangement. This means there is no first or last letter, and for a given set of letters in a given arrangement, all circular shifts of the letters are equivalent. For binary necklaces of length 4 the words 0001, 0010, 0100, 1000 are all equivalent, the words 0011, 0110, 1100, 1001 are all equivalent, the words 0101, 1010 are equivalent, and the words 0111, 1110, 1101, 1011 are all equivalent. Using the smallest word to represent each equivalence class, the 6 unique binary necklaces of length 4 are: 0000, 0001, 0011, 0101, 0111, 1111. The general question we want to answer is, given m different kinds of letters and n positions, how many unique necklaces can one construct?

First of all note that our definition of a necklace applies to many other kinds of objects. In a more commonly recognizable form, a necklace is a circular string of beads of different colors. The colors correspond to the different letters of a word. Two such necklaces that differ only by a rotation are recognized as being the same. Equivalently, a necklace can be the colored vertices or sides of a regular polygon. The following analysis applies to all these kinds of objects.

Two necklaces are equivalent if one is a circular shift of the other. We will represent these shifts as cyclic permutations. Let x_i, $i = 0, \dots, n-1$ denote the letters in a word of length n. A cyclic permutation is a circular shift of the letters given by the mapping $x_i \to x_{(i+k) \bmod n}$ for some k in the range $0 \le k \le n-1$. The set of all n cyclic permutations is called the cyclic group C_n. A set of permutations is a group if it satisfies the following conditions

- Any two permutations in succession is equivalent to a single permutation also in the set.
- The identity permutation is in the set.
- Every permutation has an inverse also in the set.
- Combining permutations is associative.

You can readily convince yourself that the set of cyclic permutations, C_n, satisfies these conditions. This is important because the method we use to count unique necklaces is based on the Polya enumeration theorem which works only for sets of permutations that form a group.

The best way to see how this works is with an example. We will take the case of $n = 6$. The permutations corresponding to circular shifts of $k = 0, 1, \dots, 5$ are shown in table 1 using Cauchy and cycle notation. In Cauchy notation, the six positions in the word are written in the top line, and the bottom line shows what positions they are shifted to by the permutation. The cycle notation shows how a set of positions cycle among themselves. For example, the cycle (1 3 5) means that position 1 shifts to 3 which shifts to 5 which shifts around to 1.

k	Cauchy	Cycle	Monomial
0	$\begin{pmatrix}1&2&3&4&5&6\\1&2&3&4&5&6\end{pmatrix}$	(1)(2)(3)(4)(5)(6)	z_1^6
1	$\begin{pmatrix}1&2&3&4&5&6\\2&3&4&5&6&1\end{pmatrix}$	(123456)	z_6
2	$\begin{pmatrix}1&2&3&4&5&6\\3&4&5&6&1&2\end{pmatrix}$	(135)(246)	z_3^2
3	$\begin{pmatrix}1&2&3&4&5&6\\4&5&6&1&2&3\end{pmatrix}$	(14)(25)(36)	z_2^3
4	$\begin{pmatrix}1&2&3&4&5&6\\5&6&1&2&3&4\end{pmatrix}$	(153)(264)	z_3^2
5	$\begin{pmatrix}1&2&3&4&5&6\\6&1&2&3&4&5\end{pmatrix}$	(165432)	z_6

Table 1: Cyclic group C_6.

Also shown in the table is the cycle index monomial for each permutation. These are constructed by using the symbol z_d for each cycle of length d and multiplying them to get the monomial for the permutation. If a permutation has two cycles of length 3 then its monomial is z_3^2. The monomials for each permutation are used to construct the cycle index polynomial[2]. It

[2]Also called simply, the cycle index

is this polynomial that will answer all of our counting questions. From it, we get what is called the pattern inventory, which is essentially a generating function for the number of necklaces with a given number of letters of each type.

The fact that we can construct a cycle index polynomial, and from that, a pattern inventory, is only possible because the set of permutations forms a group. You can construct a cycle index polynomial for any set of permutations that form a group and use it to get a pattern inventory. This general result is known as Polya's enumeration theorem. We will discuss this theorem and its generalization in a later section.

For the $n = 6$ example shown in table 1, the identity permutation breaks down into six cycles of length 1, so the cycle index monomial is z_1^6. In the next permutation, there is only one cycle of length 6, so the monomial is z_6. The $k = 2$ permutation has two cycles of length 3, so the monomial is z_3^2. The $k = 3$ permutation has three cycles of length 2, so the monomial is z_2^3. The monomials for the $k = 4$ and $k = 5$ permutations are z_3^2 and z_6 respectively. The cycle index polynomial is the sum of these monomials divided by 6, the number of permutations.

$$p(z_1, z_2, \ldots, z_6) = \frac{z_1^6 + z_2^3 + 2z_3^2 + 2z_6}{6} \qquad (14)$$

Using this procedure to find the cycle index polynomial for general values of n can be somewhat tedious. Fortunately there is a way to directly construct the polynomial without having to list all the permutations. The permutation corresponding to a particular circular shift k will be composed of cycles all of the same length. The cycle length must always be a divisor of n, since a cycle must always return to the start. If the length of the cycles in a permutation is d, a divisor of n, and the shift is k, then kd must be a multiple of n, or $k = m(n/d)$ where m is an integer less than or equal to d and relatively prime to d. So the number of k values that produce cycles of length d is equal to $\phi(d)$ which is the Euler totient function. The cycles of length d will then contribute a factor $\phi(d)z_d^{n/d}$ to the cycle index polynomial. The cycle index polynomial for C_n is then[3]

$$p(z_1, z_2, \ldots, z_n) = \frac{1}{n} \sum_{d|n} \phi(d) z_d^{n/d} \tag{15}$$

where the sum is over all the divisors of n, including 1 and n.

The simplest thing that the cycle index polynomial can tell you is how many necklaces can be made with m colors. The equation for this is found by setting $z_i = m$

[3]In the rest of the book we will often denote $p(z_1, z_2, \ldots, z_n)$ as simply $p(z)$

for all i. If you do this, equation 15 becomes

$$p_n(m) = \frac{1}{n}\sum_{d|n}\phi(d)m^{n/d} \tag{16}$$

The following is a list of these polynomials for small values of n.

$$\begin{aligned}
p_1(m) &= m \\
p_2(m) &= \frac{1}{2}(m^2 + m) \\
p_3(m) &= \frac{1}{3}(m^3 + 2m) \\
p_4(m) &= \frac{1}{4}(m^4 + m^2 + 2m) \\
p_5(m) &= \frac{1}{5}(m^5 + 4m) \\
p_6(m) &= \frac{1}{6}(m^6 + m^3 + 2m^2 + 2m)
\end{aligned} \tag{17}$$

The general formula for when n is a prime number is

$$p_n(m) = \frac{1}{n}(m^n + (n-1)m) \tag{18}$$

The cycle index polynomial can provide a lot more information than just the total number of necklaces. It can also be used to determine how many necklaces there are with a given number of beads of each color. This is called the pattern inventory. In the simplest case, suppose we want to construct necklaces with beads of just two colors, represented by the symbols r (red), and g (green). In the cycle index polynomial, we make the following substitution for z_d.

$$z_d = r^d + g^d \tag{19}$$

Looking at the example of $n = 4$, the cycle index polynomial is

$$p(z_1, z_2, z_3, z_4) = \frac{1}{4}(z_1^4 + z_2^2 + 2z_4) \tag{20}$$

Now making the above substitution, multiplying out, and simplifying, we get

$$p(r, g, b) = r^4 + gr^3 + 2g^2r^2 + g^3r + g^4 \tag{21}$$

The equation says there is one all red, and one all green necklace, one necklace with three red beads and one green, one necklace with three green beads and one red, and two necklaces with two red and two green beads. The general idea is that the coefficient of the monomial $g^k r^j$ is the number of necklaces with k green and j red beads where $k + j = n$, the total number of beads, or length of the necklace.

To get the pattern inventory for more colors, additional symbols for each color must be added to the substitution in equation 19. For more examples, see the problems section.

COUNTING BRACELETS

A bracelet is like a necklace except that now there are more symmetries. Instead of just rotational symmetry, about an axis perpendicular to the plane of the necklace, we have 180° rotational symmetries about axes in the plane of the necklace. In other words, a bracelet is a necklace that can be picked up and turned over, as well as rotated. Two bracelets are equivalent if one can be transformed into another by rotation and/or turning over. In terms of strings, if one string can be gotten from another by a circular shift and/or a reversal of the string, then the two strings are equivalent.

For example there are 11 necklaces of length 3 that use 3 symbols (colors), they are: 000, 001, 002, 011, 012, 021, 022, 111, 112, 122, 222. Take the pair of strings 012 and 021 from this list. A circular shift of the first string to the left one position produces 120 and reversing this turns it into the second string, 021. These two strings represent the same bracelet. No other pair of strings in the list can be related in this way so the number of bracelets of length 3 in 3 symbols is 10.

The group of permutations that define bracelets of length n include the permutations of the cyclic group C_n plus the permutations that correspond to rotation and reversal. These latter permutations show the way the vertices of a regular n sided polygon are permuted by reflections across lines in the plane of the polygon. As an example, look at the two cases where $n = 3$ and $n = 4$, shown in the figure below.

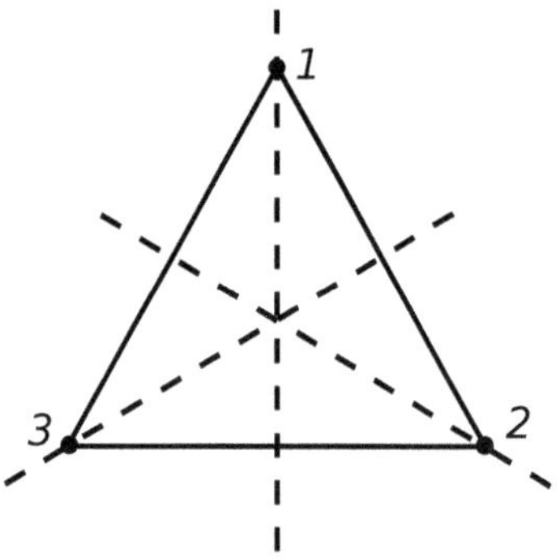

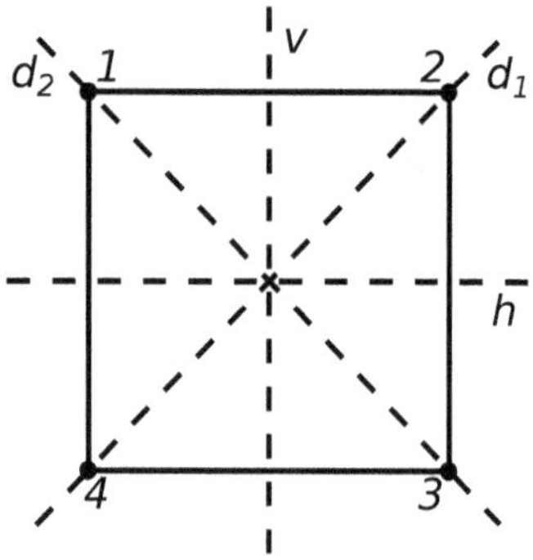

For $n = 3$ we are permuting the vertices of an equilateral triangle and there are three possible reflection lines. Each line goes through a vertex and bisects the opposite side. A reflection across the line through vertex 1 will exchange vertices 2 and 3 and keep vertex 1 fixed, so in cycle notation the permutation is $(1)(23)$. Similar reflections about the lines through vertices 2 and 3 will result in the permutations $(2)(13)$ and $(3)(12)$ respectively. Table 2 shows the complete set of permutations for the $n = 3$ bracelet. The first three correspond to rotations of the triangle by $0°$, $120°$, and $240°$. The last three correspond to the reflections.

For $n = 4$ we are permuting the vertices of a square and we have four reflection lines. Two of the lines bisect

$0°$	(1)(2)(3)	z_1^3
$120°$	(123)	z_3
$240°$	(132)	z_3
l_1	(1)(23)	$z_1 z_2$
l_2	(2)(13)	$z_1 z_2$
l_3	(3)(12)	$z_1 z_2$

Table 2: Dihedral group D_3.

opposite sides of the square and two join vertices along the diagonals of the square. A reflection across the vertical line produces the permutation (12)(34). A reflection across the horizontal line produces the permutation (14)(23). Reflections across the diagonals d_1 and d_2 produce the permutations (1)(3)(24) and (13)(2)(4) respectively. Table 3 shows the complete set of permutations for the $n = 4$ bracelet. The first four correspond to rotations of the square by $0°$, $90°$, $180°$, and $270°$. The last four correspond to the reflections.

$0°$	(1)(2)(3)(4)	z_1^4
$90°$	(1234)	z_4
$180°$	(13)(24)	z_2^2
$270°$	(1432)	z_4
v	(12)(34)	z_2^2
h	(14)(23)	z_2^2
d_1	(1)(3)(24)	$z_1^2 z_2$
d_2	(2)(4)(13)	$z_1^2 z_2$

Table 3: Dihedral group D_4.

In general, for a bracelet of length n, there are always n permutations due to rotations and n permutations due to reflections. These $2n$ permutations form what is called the dihedral group D_n.

Also shown in the tables for $n = 3$ and $n = 4$ is the cycle index monomial for each permutation. They are constructed the same way as for necklaces, each cycle of length d is replaced by the symbol z_d. The cycle index polynomial for $n = 3$ bracelets is then

$$P(z) = \frac{1}{6}(z_1^3 + 2z_3 + 3z_1z_2) \tag{22}$$

and the cycle index polynomial for $n = 4$ bracelets is

$$P(z) = \frac{1}{8}(z_1^4 + z_2^2 + 2z_4 + 2z_2^2 + 2z_1^2z_2) \tag{23}$$

As with necklaces, it is possible to directly construct the cycle index polynomial for bracelets of length n without having to list all the permutations. The part due to rotations will be the same as for the necklaces. When n is odd, the contribution due to the reflections will be

$$nz_1z_2^{(n-1)/2} \tag{24}$$

and when n is even, the contribution is

$$\frac{n}{2}(z_1^2 z_2^{(n-2)/2} + z_2^{n/2}) \tag{25}$$

Putting it all together, the cycle index polynomial for bracelets is

$$p(z) = \frac{1}{2n}\sum_{d|n} \phi(d) z_d^{n/d} + \begin{cases} \frac{1}{2} z_1 z_2^{(n-1)/2}, & n = \text{odd} \\ \frac{1}{4}(z_1^2 z_2^{(n-2)/2} + z_2^{n/2}), & n = \text{even} \end{cases} \tag{26}$$

where the sum is over all the divisors of n, including 1 and n.

To find the number of bracelets that can be made with m colors, set $z_i = m$ for all i. If you do this, equation 26 becomes

$$p_n(m) = \frac{1}{2n}\sum_{d|n} \phi(d) m^{n/d} + \begin{cases} \frac{1}{2} m^{(n+1)/2}, & n = \text{odd} \\ \frac{1}{4}(m^{(n+2)/2} + m^{n/2}), & n = \text{even} \end{cases} \tag{27}$$

The following is a list of these polynomials for small values of n.

$$\begin{aligned}
p_3(m) &= \frac{1}{6}(m^3 + 3m^2 + 2m) \qquad (28)\\
p_4(m) &= \frac{1}{8}(m^4 + 2m^3 + 3m^2 + 2m)\\
p_5(m) &= \frac{1}{10}(m^5 + 5m^3 + 4m)\\
p_6(m) &= \frac{1}{12}(m^6 + m^4 + 4m^3 + 2m^2 + 2m)\\
p_7(m) &= \frac{1}{14}(m^7 + 7m^4 + 6m)\\
p_8(m) &= \frac{1}{16}(m^8 + 4m^5 + 5m^4 + 2m^2 + 4m)
\end{aligned}$$

As with necklaces, the cycle index polynomial can tell you how many bracelets there are with a given number of beads of a certain color. Suppose for example, we want to find the number of bracelets of length 4, composed of beads colored red, green or blue, that have exactly 2 red beads. If we represent the colors with the symbols r (red), g (green) and b (blue), then one way to do this would be to make the following substitution for z_d in the cycle index polynomial

$$z_d = r^d + g^d + b^d \qquad (29)$$

but since we're only interested in counting red beads, we can set $g = b = 1$ and use the simpler substitution

$z_d = r^d + 2$. Doing this and simplifying, we get

$$p(r) = r^4 + 2r^3 + 6r^2 + 6r + 6 \tag{30}$$

The number of bracelets with exactly 2 red beads is 6, the coefficient of r^2 in $p(r)$. In general, the coefficient of r^k in $p(r)$ is equal to the number of bracelets that have k red beads. So there are 6 bracelets with no red beads, 6 with only one, 6 with two, 2 with three, and 1 with four. For more examples, see the problems section.

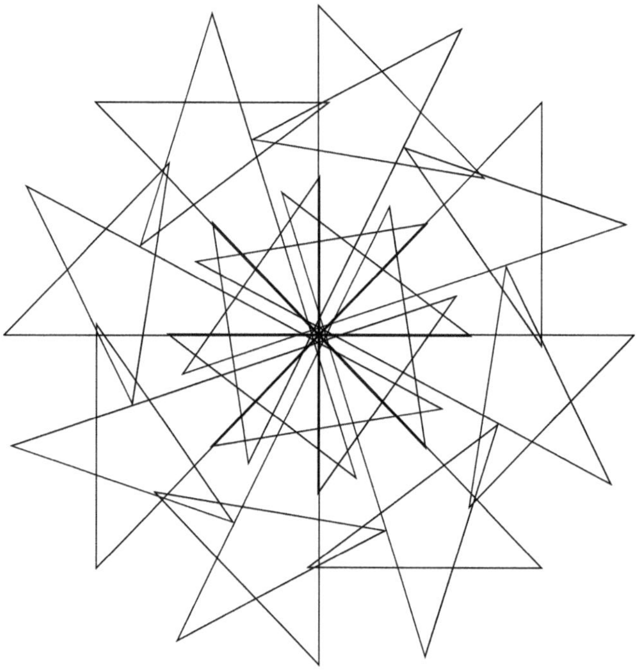

POLYA AND BURNSIDE THEOREMS

The method we used to count necklaces and bracelets in the previous two sections are particular examples of Polya's Theory of Counting. This theory is used to count the number of unique objects in a set of objects when there is equivalence due to symmetry. The symmetries are expressed in terms of permutations that turn one object into another. Two objects that are related by a permutation are equivalent and counted as one. The theory only works when the set of permutations forms a permutation group.

As a simple example lets look at the set of three bit binary strings. There are of course eight of them: 000, 001, 010, 011, 100, 101, 110, 111. Now lets impose the condition that two strings are equivalent if one is the reverse of the other. The pairs of strings 001, 100 and 011, 110 are then equivalent. There are only 6 three bit binary strings under reversal symmetry, they are: 000, 001, 010, 011, 101, 111[4].

There is only one symmetry in this case, reversal, which is its own inverse so the permutation group is $G = \{(1)(2)(3), (13)(2)\}$. The first permutation is the iden-

[4]Note that when a set of strings is equivalent we use the string that is smallest in lexicographic order to represent the set

tity which leaves all three bits alone. The second permutation reverses the string. The first and last bits are swapped and the middle bit is left alone.

Polya's theory says we can get all the information we need from the permutation group by constructing what's called the cycle index polynomial. For each permutation, replace each cycle of length d with the variable z_d. The permutation (1)(2)(3) becomes z_1^3 and (13)(2) becomes $z_1 z_2$. Adding these terms and dividing by the size of the group gives us the cycle index polynomial

$$p(z) = \frac{1}{2}(z_1^3 + z_1 z_2) \tag{31}$$

Now let x and y represent 1 and 0 respectively and make the substitution $z_d = x^d + y^d$ in the cycle index. Simplifying, we get

$$p(x, y) = x^3 + 2x^2y + 2xy^2 + y^3 \tag{32}$$

The coefficient of $x^j y^k$ in this expression is the number of strings with j 1's and k 0's. Comparing, we see that this matches the six strings listed above. To get the total number of strings, set $x = 1$, $y = 1$ in equation 32 or equivalently set $z_d = 2$ in equation 31.

So the general procedure is to first determine the permutation group.[5] Then for each permutation, replace each cycle of length d with the variable z_d. Sum up

[5]The procedure only works if the set of permutations forms a group.

the terms and divide by the number of permutations to get the cycle index polynomial. To find the number of unique objects that can be produced using m symbols (colors), set $z_d = m$ for all d in the cycle index polynomial and evaluate the result.

To find the number of unique objects where each symbol (color) appears a given number of times, let x_i represent the ith symbol (color) and set

$$z_d = x_1^d + x_2^d + \cdots + x_m^d \tag{33}$$

for all d in the cycle index polynomial. The coefficient of $x_1^{l_1} x_2^{l_2} \cdots x_m^{l_m}$ in the result is the number of objects where the ith symbol (color) appears l_i times.

There is a generalization of Polya's theorem that can be used in cases where there are symmetries not just of the positions of symbols in a word but also of the values of the symbols themselves. For example we may want to count the number of reversible strings of three symbols where we have equivalence with respect to all permutations of the values of the symbols. Using the symbols $\{0, 1, 2\}$, the length 2 strings: 00 11 22 are all equivalent, and the strings: 01 10 02 20 12 21 are all equivalent. There are only two unique length 2 strings: 00 01.

The way to solve these kinds of problems is to identify the permutation group for the strings and the permutation group for the symbols. Then create the cycle index polynomial for each group. Let the cycle index for the strings be $p_G(z)$ and the cycle index for the symbols be $p_H(z)$, where we have used z to represent the set of variables $z_1, z_2, \ldots$. Next make the following substitution in $p_G(z)$, the string cycle index,

$$z_i^n \to \frac{\partial^n}{\partial z_i^n}$$

and make the following substitution in $p_H(z)$, the symbol cycle index,

$$\begin{aligned} z_i &\to e^{i\Sigma_i} \\ \Sigma_i &= \sum_{k=1}^{\infty} z_{i\cdot k} \end{aligned}$$

Finally operate on p_H with p_G and evaluate the result at $z_i = 0$ for all i. Several examples of using this method can be found in the problems section. For more of an explanation of how and why this method works see the book by Liu in the references section.

Polya's Theorem works when there is the same choice for all the elements of a pattern. For example, each bead of a necklace can be red, green, or blue, or each letter of a word can be a 0 or 1. There are problems, however, for which this is not the case. If we want words where two or more 1's do not appear adjacent, then if one letter is a 1, the next must be a 0. For these kinds of problems, we can use Burnside's Theorem, which says that the number of unique objects under a permutation group G is given by

$$\frac{1}{|G|}\sum_{g\in G}\psi(g) \tag{34}$$

where $\psi(g)$ is the number of objects that are invariant under the permutation g.

For example, suppose we want the number of unique binary strings of length $2n$ with an equal number of 1's and 0's under the symmetry of reversal. In other words, two strings are considered equivalent if one is the reverse of the other. The permutation group has two elements, the identity g_0, and the element g_1 that reverses the string. All strings are invariant under the identity, so $\psi(g_0)$ is simply the number of strings of length $2n$ with equal numbers of 1's and 0's, $\psi(g_0) = \binom{2n}{n}$. A string invariant under reversal will look like $b_1b_2\ldots b_nb_n\ldots b_2b_1$. If the string is to have the same number of 1's and 0's then the first half of the string, $b_1b_2\ldots b_n$ must also have the same number of 1's and 0's. This is only possible if n is even. Therefore,

$$\psi(g_1) = \begin{cases}\binom{n}{n/2}, & n = \text{even}\\ 0, & n = \text{odd}\end{cases} \tag{35}$$

Putting $\psi(g_0)$ and $\psi(g_1)$ into equation 34 we get the following formula for the number of binary strings of length $2n$ with an equal number of 1's and 0's under the symmetry of reversal.

$$c(n) = \frac{1}{2} \begin{cases} \binom{2n}{n} + \binom{n}{n/2}, & n = \text{even} \\ \binom{2n}{n}, & n = \text{odd} \end{cases} \tag{36}$$

Burnside's Theorem must also be used for counting necklaces where there is a restriction in the way the beads can be colored. Suppose we want to count the number of binary necklaces where two or more black beads do not appear adjacent. Call these Fibonacci necklaces. This can only be done using Burnside's Theorem. The group of symmetries is the cyclic group C_n composed of the elements g_i, $i = 0, 1, \ldots, n-1$ where g_i is a rotation by i positions (beads). Equation 34 says that the number of unique necklaces is

$$c(n) = \frac{1}{n} \sum_{i=0}^{n-1} \psi(g_i) \tag{37}$$

where $\psi(g_i)$ is the number of necklaces invariant under a rotation by i positions. For such an invariance to be possible, the necklace must be composed of repeats of a circular Fibonacci string of length i. A Fibonacci string is a binary string where two or more 1's never appear adjacent. A circular Fibonacci string is a Fibonacci string where the two ends are considered adjacent.

For example, with $n = 6$, necklaces invariant under g_3 must look like $b_1b_2b_3b_1b_2b_3$ where $b_1b_2b_3$ is a circular Fibonacci string of length 3. There are four such strings. They are 000, 001, 010 and 100. Note that 101 is not admissible because b_1 is adjacent to b_3 in a circular string. Therefore, we have $\psi(g_3) = 4$. In general, if i is a divisor of n then $\psi(g_i) = f(i)$ where $f(i)$ is the number of circular Fibonacci strings of length i, otherwise $\psi(g_i) = f(1)$.

Equation 37 can be simplified somewhat. Each permutation g_i of the cyclic group C_n will be composed of cycles of equal length. The cycle lengths must be divisors of n. If d is a divisor of n then there will be permutations with n/d cycles of length d, and there are $\phi(d)$ such permutations (see the section on necklaces). A necklace is invariant under such a permutation if it is composed of circular Fibonacci words of length n/d therefor the number of Fibonacci necklaces is

$$\frac{1}{n}\sum_{d|n}\phi(d)f(n/d) \tag{38}$$

This equation is valid for other kinds of pattern restrictions. Let $f(n)$ be the number of circular strings with a given pattern restriction, then the formula will give the number of necklaces with that pattern restriction.

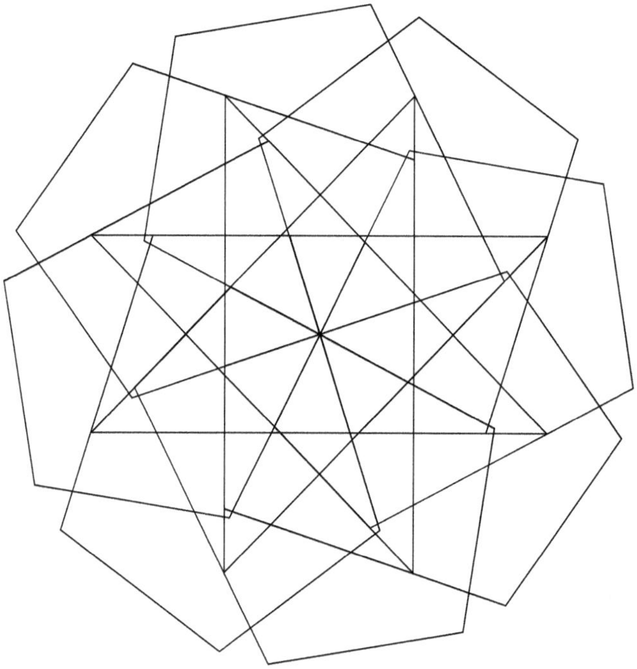

PROBLEMS AND SOLUTIONS

Problem 1. Using an alphabet of m letters, how many strings of length n are possible?

Solution. We have no restrictions and no symmetries, so there are m choices for each of the n letters, and the total number of strings is m^n.

Another way to look at this problem is that we do have one symmetry, which is the identity. The identity permutation is $(1)(2)\cdots(n)$, i.e. there are n cycles of length 1. The cycle index polynomial is then

$$p(z) = z_1^n \tag{39}$$

Making the substitution $z_1 \to m$ we get $p(m) = m^n$ for the number of strings of length n using m letters.

Problem 2. In the previous problem, suppose we are using two letters, a and b. How many strings of length n are there, where the letter a appears exactly k times?

Solution. This is a simple combinatorial problem. We want to choose k out of the n letters to be a's. Since the order is not important, the number of ways to do it is $\binom{n}{k}$.

The problem can also be solved using the cycle index polynomial in equation 39. Make the substitution $z_1 \to a + b$ to get $p(a, b) = (a + b)^n$. The answer is the coefficient of $a^k b^{n-k}$ in the expansion of $(a+b)^n$ which is $\binom{n}{k}$. Since there are only two letters, we can simplify things a bit and use $z_1 \to a+1$. Now, the answer is the coefficient of a^k in the expansion of $(a+1)^n$ which is once again $\binom{n}{k}$.

Problem 3. Using an alphabet of three letters, a, b and c, how many strings of length n are there that contain one a and one b?

Solution. We will solve this problem in three different ways. First the straightforward combinatorial approach. We need to select two of the n letters to be a and b. This can be done in $\binom{n}{2}$ ways, and for each of these ways there are two ways to place the a and the b. So the answer is $2\binom{n}{2} = n(n-1)$.

The second solution involves using the cycle index polynomial in equation 39. Make the substitution $z_1 \to a+b+c$ to get $p(a,b,c) = (a+b+c)^n$. The answer is the coefficient of abc^{n-2} in the expansion, which is $\binom{n}{1,1} = n(n-1)$.

For the third solution we will use a regular expression to find the generating function for the number of strings. The regular expression that matches the strings we want is $c^*(ac^*b + bc^*a)c^*$. In words, the expression says we want strings that begin with any number of c's followed by an a and b or b and a with any number of c's in between, and ending with any number of c's. To get the generating function, make the substitutions $a \to z$, $b \to z$, and $c^* \to 1/(1-z)$. The generating function is

$$G(z) = \frac{1}{1-z}\left(\frac{2z^2}{1-z}\right)\frac{1}{1-z} = \frac{2z^2}{(1-z)^3} \tag{40}$$

The Taylor series expansion of $G(z)$ is

$$G(z) = 2z^2 + 6z^3 + 12z^4 + 20z^5 + 30z^6 + 42z^7 + 56z^8 + \cdots \tag{41}$$

The coefficient of z^n in the expansion is $n(n-1)$ which is our answer. The integer sequence is A002378 in the On-Line Encyclopedia of Integer Sequences (OEIS).

Problem 4. We want to tile an $n \times 1$ grid with 1×1 black and white tiles. How many ways can this be done so that two or more black tiles do not appear side by side?

Solution. Let's represent the layout of the tiles as an n bit binary number, where 1 represents a black tile, and 0 a white tile. We can have a run of zero or one 1's followed by a 0, and this can be repeated any number of times. A final run of zero or one 1's can appear at the end. A regular expression that matches these binary strings is

$$R = ((\epsilon + 1)0)^*(\epsilon + 1) \tag{42}$$

With the regular expression we can get the generating function for the number of strings. In the regular expression, make the substitutions $0 \to z$, $1 \to z$, $\epsilon \to 1$, and the starred expression $f(z) = (1+z)z$ becomes $1/(1-f(z))$. This produces the generating function

$$G(z) = \frac{1+z}{1-z-z^2} \tag{43}$$

The Taylor series expansion of $G(z)$ is

$$G(z) = 1+2z+3z^2+5z^3+8z^4+13z^5+21z^6+34z^7+\ldots \tag{44}$$

The integer sequence is A000045 in the OEIS. The numbers are the well known Fibonacci numbers.

We can also solve the problem via a recurrence relation. Note that every tiling must end with either a white tile, or a white tile followed by a single black tile. This means there are two ways to make a tiling of length n. Add a white tile to the end of a tiling of length $n-1$. Add a white tile followed by a black tile to the end of a tiling of length $n-2$. Therefore, if a_n is the number of tilings of length n, then it must be that

$$a_n = a_{n-1} + a_{n-2} \tag{45}$$

This recurrence relation can be iterated to find a_n given the two starting values $a_0 = 1, a_1 = 2$. We have, $a_2 = 1+2 = 3$, $a_3 = 2+3 = 5$, $a_4 = 3+5 = 8$, and so on.

All the tilings for $n = 8$ are shown below (there are 55).

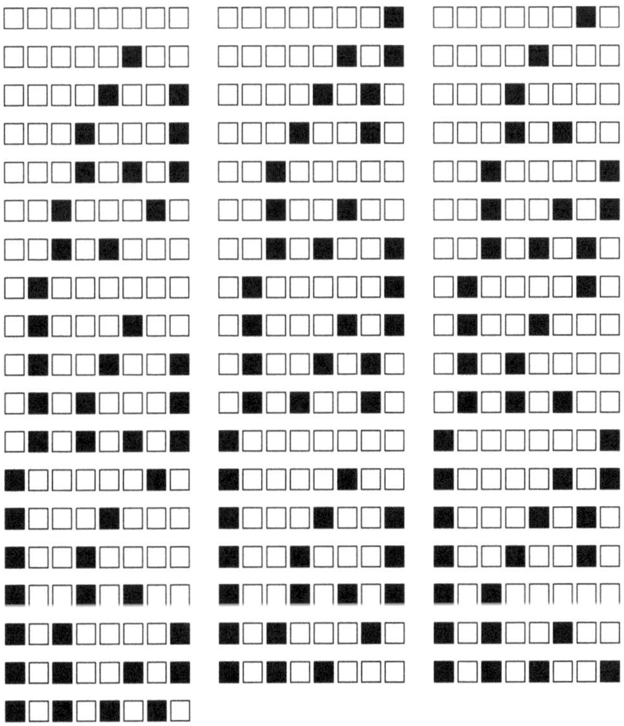

The tilings can be generated by the following automaton where a 0 represents a white square and a 1 a black square.

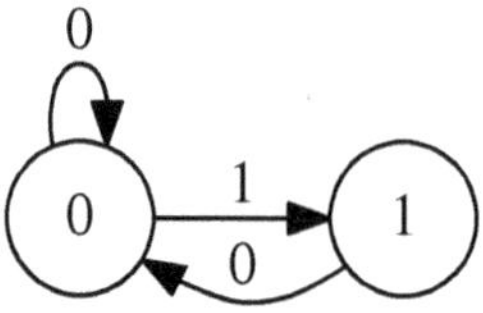

The command for generating the example above is:

```
autogen fibo.aut 8 0 0 1
```

where `fibo.aut` is a file representing the automaton. It consists of the following lines:

```
2
0 00 11
1 00
```

For an explanation of the meaning of these lines, and the use of the **`autogen`** program, see the appendix.

Problem 5. Repeat the previous problem, but now we want the number of tilings where two or more black tiles do appear side by side.

Solution. Once again, represent the layout of the tiles as an n bit binary number, where 1 represents a black tile, and 0 a white tile. In the previous problem, we found the generating function for the number of n bit binary numbers that do not contain a run of two or more 1's. Now we want the generating function for the number of binary numbers that do contain such a run. A binary number either does or does not contain a run of two or more 1's, so the sum of the *contain* and *not-contain* generating functions should equal the generating function for the number of n bit binary numbers, which is

$$\frac{1}{1-2z} = 1 + 2z + 4z^2 + 8z^3 + 16z^4 + \cdots \tag{46}$$

Now subtracting equation 43 from this, we get the generating function we want.

$$G(z) = \frac{z^2}{1 - 3z + z^2 + 2z^3} \tag{47}$$

The Taylor series expansion of $G(z)$ is

$$G(z) = z^2 + 3z^3 + 8z^4 + 19z^5 + 43z^6 + 94z^7 + 201z^8 + \ldots \tag{48}$$

The integer sequence is A008466 in the OEIS.

We could also solve the problem directly by constructing a regular expression that matches the binary strings that we want. The regular expression is

$$R = ((\epsilon + 1)0)^*11(0+1)^* \tag{49}$$

Making the substitutions like we did in the previous problem turns this into the generating function in equation 47.

All the tilings for $n = 6$ are shown below (there are 43).

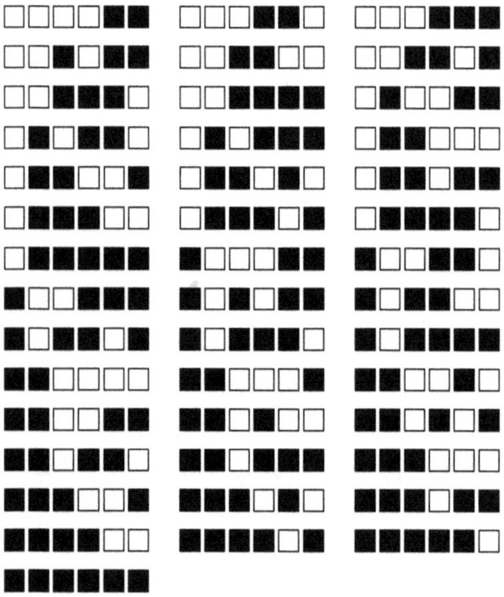

The tilings can be generated by the following automaton where a 0 represents a white square and a 1 a black square.

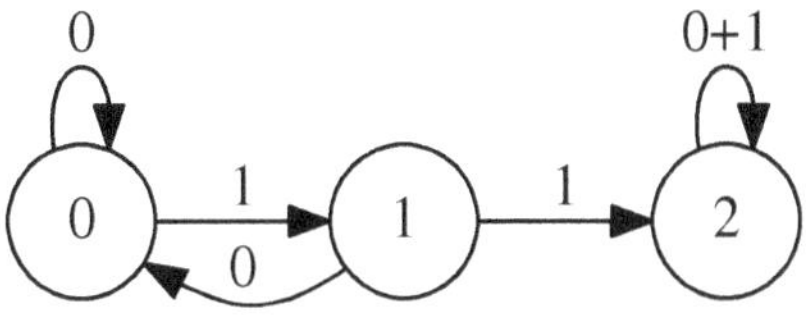

The command for generating the example above is:

```
autogen fibonot.aut 6 0 2
```

where `fibonot.aut` defines the automaton, consisting of the following lines:

```
3
0 00 11
1 00 12
2 02 12
```

Problem 6. This is like the previous two tiling problems, but now we want the number of tilings where a black tile is followed by a white tile at least once.

Solution. Using 1 to represent a black tile, and 0 a white tile, we can write the following regular expression to match the tiling patterns we want.

$$R = 0^*11^*0(0+1)^* \qquad (50)$$

The regular expression matches any number of 0's followed by a least one 1 and a 0 followed by any pattern of 0's and 1's. This is converted into a generating function using the same substitutions as in the previous two problems. The generating function is

$$G(z) = \frac{z^2}{(1-z)^2(1-2z)} \tag{51}$$

The Taylor series expansion of $G(z)$ is

$$G(z) = z^2 + 4z^3 + 11z^4 + 26z^5 + 57z^6 + 120z^7 + 247z^8 + \dots \tag{52}$$

The integer sequence is A000295 in the OEIS. The numbers are called the Eulerian numbers.

All the tilings for $n = 6$ are shown below (there are 57).

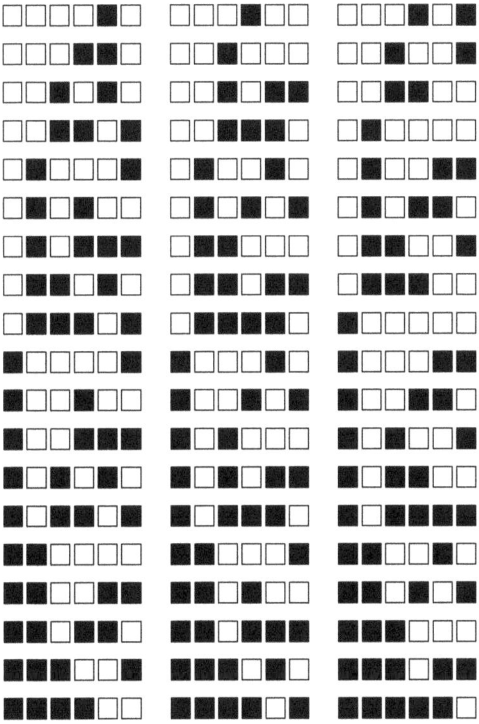

The tilings can be generated by the following automaton where a 0 represents a white square and a 1 a black square.

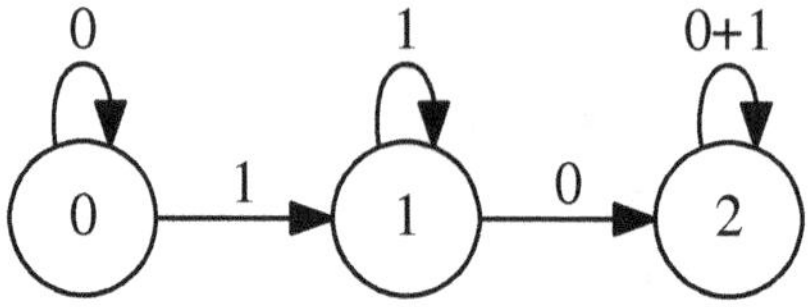

The command for generating the example above is:

```
autogen eulerian.aut 6 0 2
```

where `eulerian.aut` defines the automaton, consisting of the following lines:

```
3
0 00 11
1 02 11
2 02 12
```

Problem 7. Same as the previous problem, but now we want the number of tilings where a black tile is never followed by a white tile.

Solution. To get the generating function, we subtract the generating function for the previous problem, given by equation 51, from the generating function for the number of n bit binary numbers given in equation 46. The generating function is

$$G(z) = \frac{1}{(1-z)^2} \tag{53}$$

The Taylor series expansion of $G(z)$ is

$$G(z) = 1+2z+3z^2+4z^3+5z^4+6z^5+7z^6+8z^7+9z^8+\dots \tag{54}$$

These are simply the natural numbers.

All the tilings for $n = 8$ are shown below (there are 9).

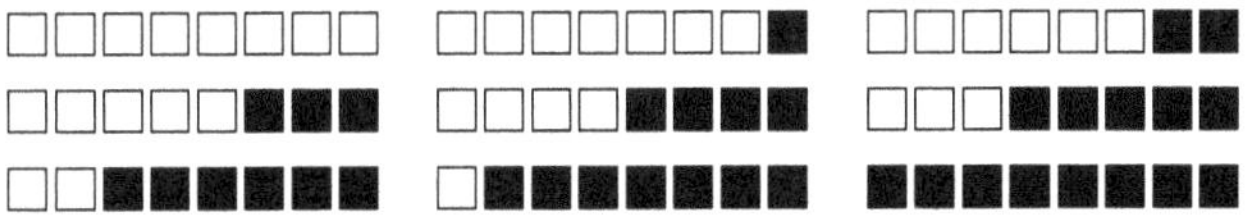

The command for generating these tilings is:

```
autogen eulerian.aut 8 0 0 1
```

Problem 8. We want to tile an $n \times 1$ grid with 1×1 black and white tiles. How many ways can this be done so that three or more black tiles do not appear side by side?

Solution. Let's represent the layout of the tiles as an n bit binary number, where 1 represents a black tile, and 0 a white tile. We can have a run of zero, one, or two 1's followed by a 0, and this can be repeated any number of times. A final run of zero, one, or two 1's can appear at the end. A regular expression that matches these binary strings is

$$R = ((\epsilon + 1 + 11)0)^*(\epsilon + 1 + 11) \tag{55}$$

With the regular expression we can get the generating function for the number of strings. In the regular expression, make the substitutions $0 \to z$, $1 \to z$, $\epsilon \to 1$, and the starred expression $f(z) = (1+z+z^2)z$ becomes $1/(1 - f(z))$. This produces the generating function

$$G(z) = \frac{1 + z + z^2}{1 - z - z^2 - z^3} \tag{56}$$

The Taylor series expansion of $G(z)$ is

$$G(z) = 1+2z+4z^2+7z^3+13z^4+24z^5+44z^6+81z^7+\dots \tag{57}$$

The integer sequence is A000073 in the OEIS. The numbers are called the tribonacci numbers.

We can also solve the problem via a recurrence relation. Note that every tiling must end with either a white tile, a white tile followed by a single black tile, or a white tile followed by two black tiles. This means there are three ways to make a tiling of length n. Add a white tile to the end of a tiling of length $n - 1$. Add a white tile followed by a black tile to the end of a tiling of length $n-2$. Add a white tile followed by two black tiles to the end of a tiling of length $n-3$. Therefore, if a_n is the number of tilings of length n, then it must be that

$$a_n = a_{n-1} + a_{n-2} + a_{n-3} \tag{58}$$

This recurrence relation can be iterated to find a_n given the three starting values $a_0 = 1, a_1 = 2, a_2 = 4$. We have, $a_3 = 1 + 2 + 4 = 7$, $a_4 = 2 + 4 + 7 = 13$, $a_5 = 4 + 7 + 13 = 24$, and so on.

All the tilings for $n = 6$ are shown below (there are 44).

The command for generating these tilings is:

```
autogen tribo.aut 6 0 0 1 2
```

where `tribo.aut` is the automaton file consisting of the following lines:

```
3
0 00 11
1 00 12
2 00
```

Problem 9. How many n bit binary numbers contain the three bit pattern 111?

Solution. The appearance of 111 in a binary string can be represented by the automaton in the figure below, with start state **0** and end state **E**.

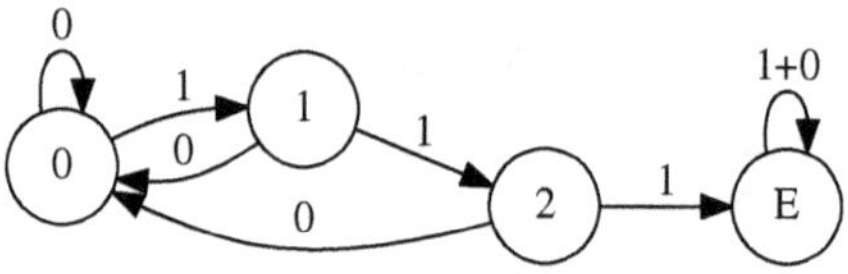

To get the regular expression, we simplify the automaton by eliminating state **1**, then state **2** as shown below

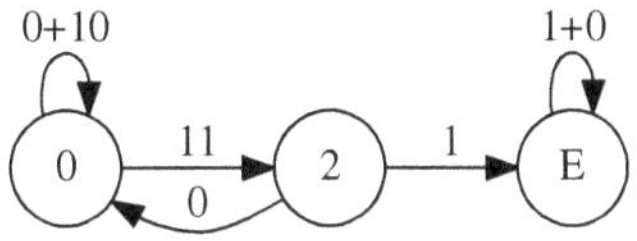

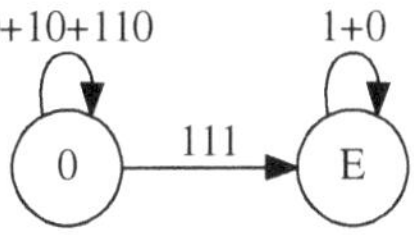

From the simplified automaton, you can read off the regular expression as follows

$$R = (0 + 10 + 110)^{*}111(1 + 0)^{*} \tag{59}$$

To get the generating function, make the substitutions $0 \to z$, $1 \to z$, and $f(z)^{*} \to 1/(1 - f(z))$.

$$G(z) = \frac{z^3}{(1 - z - z^2 - z^3)(1 - 2z)} \tag{60}$$

The Taylor series expansion of $G(z)$ is

$$G(z) = z^3 + 3z^4 + 8z^5 + 20z^6 + 47z^7 + 107z^8 + 238z^9 + \cdots \tag{61}$$

The values for strings of length 1 through 10 are

n	1	2	3	4	5	6	7	8	9	10
$f(n)$	0	0	1	3	8	20	47	107	238	520

The integer sequence is A050231 in the OEIS. This sequence is also that associated with the appearance of pattern 000, being the bitwise-not of pattern 111.

All the binary numbers for $n = 7$ are shown below (there are 47).

0000111	0001110	0001111	0010111	0011100
0011101	0011110	0011111	0100111	0101110
0101111	0110111	0111000	0111001	0111010
0111011	0111100	0111101	0111110	0111111
1000111	1001110	1001111	1010111	1011100
1011101	1011110	1011111	1100111	1101110
1101111	1110000	1110001	1110010	1110011
1110100	1110101	1110110	1110111	1111000
1111001	1111010	1111011	1111100	1111101
1111110	1111111			

The command for generating these numbers is:

```
autogen tribonot.aut 7 0 E
```

where `tribonot.aut` is the automaton file consisting of the following lines:

```
4
0 00 11
1 00 12
2 00 1E
E 0E 1E
```

Problem 10. How many n bit binary numbers do not contain the three bit pattern 111?

Solution. This is essentially the same question as Problem 8, but let's get the regular expression with an automaton. The avoidance of 111 in a binary string can be represented by the automaton in the figure below, with start state **0** and end state **E**.

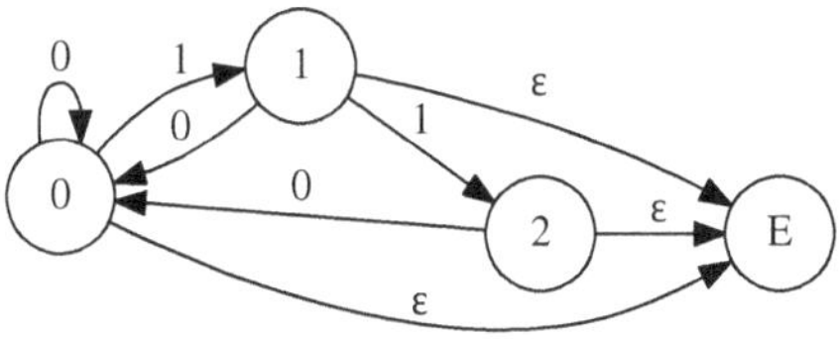

Simplifying the automaton as in the last problem, we get

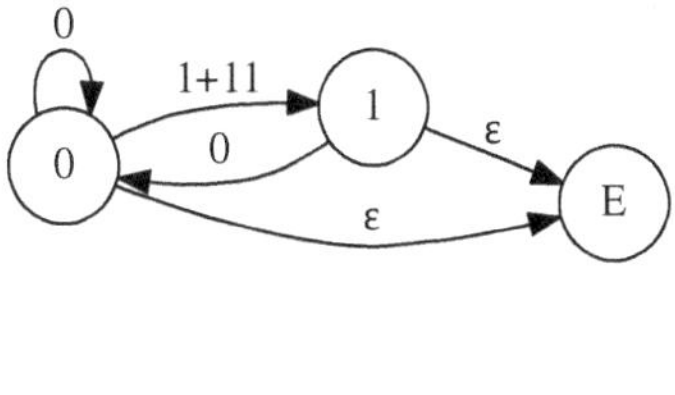

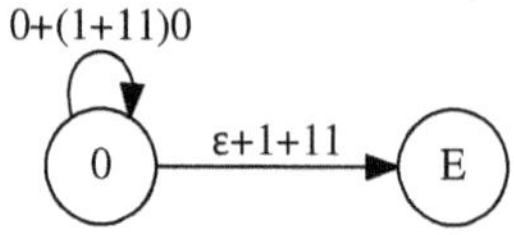

From the simplified automaton, the regular expression is

$$\begin{aligned} R &= (0+(1+11)0)^*(\epsilon+1+11) \\ &= (0+10+110)^*(\epsilon+1+11) \end{aligned} \quad (62)$$

which is the same regular expression as that of Problem 8, so the solution is identical.

Problem 11. How many n bit binary numbers contain the three bit pattern 110?

Solution. The appearance of 110 in a binary string can be represented by the automaton in the figure below, with start state **0** and end state **E**.

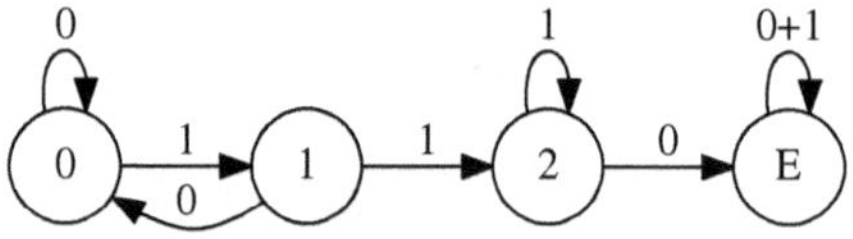

Simplifying as before

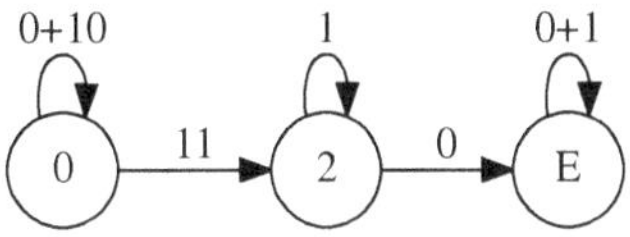

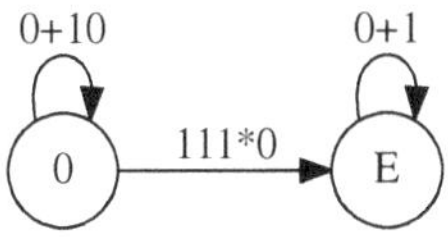

The regular expression is then

$$R = (0 + 10)^*111^*0(0 + 1)^* \tag{63}$$

With the usual substitutions, the generating function is

$$G(z) = \frac{z^3}{(1 - z - z^2)(1 - z)(1 - 2z)} \tag{64}$$

The Taylor series expansion of $G(z)$ is

$$G(z) = z^3 + 4z^4 + 12z^5 + 31z^6 + 74z^7 + 168z^8 + 369z^9 + \cdots \tag{65}$$

The values for strings of length 1 through 10 are

n	1	2	3	4	5	6	7	8	9	10
$f(n)$	0	0	1	4	12	31	74	168	369	792

The integer sequence is A232580 in the OEIS. This sequence is also that associated with the appearance of patterns 011, 001 and 100, being the reverse, the bitwise-not, and the bitwise-not of the reverse, of pattern 110.

All the binary numbers for $n = 6$ are shown below (there are 31).

```
000110 001100 001101 001110 010110
011000 011001 011010 011011 011100
011101 011110 100110 101100 101101
101110 110000 110001 110010 110011
110100 110101 110110 110111 111000
111001 111010 111011 111100 111101
111110
```

The command for generating these numbers is:

```
autogen 110.aut 6 0 E
```

where `110.aut` is the automaton file consisting of the following lines:

```
4
0 00 11
1 00 12
2 0E 12
E 0E 1E
```

Problem 12. How many n bit binary numbers do not contain the three bit pattern 110?

Solution. The avoidance of 110 in a binary string can be represented by the automaton in the figure below, with start state **0** and end state **E**.

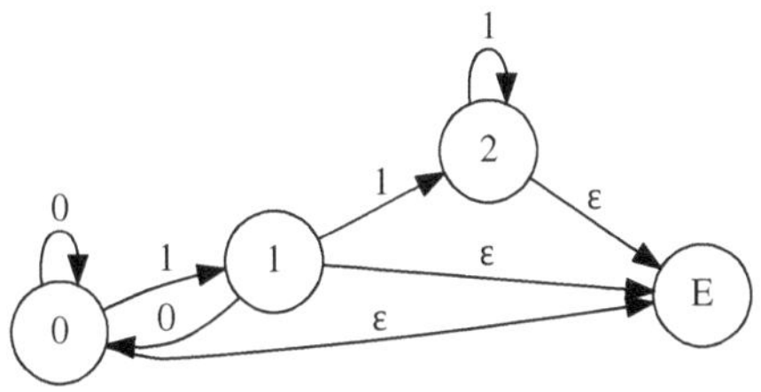

Simplifying as before

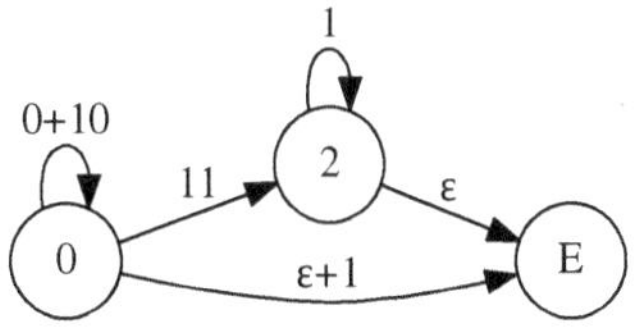

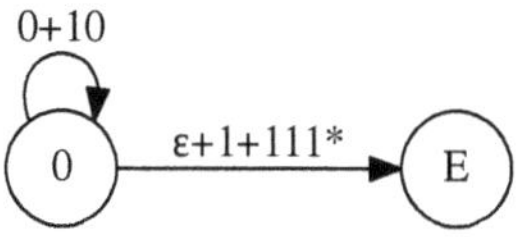

The regular expression is then

$$R = (0+10)^*(\epsilon + 1 + 111^*) \tag{66}$$

The generating function is

$$G(z) = \frac{1}{(1-z-z^2)(1-z)} \tag{67}$$

The Taylor series expansion of $G(z)$ is

$$G(z) = 1+2z+4z^2+7z^3+12z^4+20z^5+33z^6+54z^7+88z^8+\cdots \tag{68}$$

The values for strings of length 1 through 10 are

n	1	2	3	4	5	6	7	8	9	10
$f(n)$	2	4	7	12	20	33	54	88	143	232

The integer sequence is A000071 in the OEIS. It is the Fibonacci numbers minus 1. This sequence is also that associated with the avoidance of patterns 011, 001 and 100, being the reverse, the bitwise-not, and the bitwise-not of the reverse, of pattern 110.

All the binary numbers for $n = 6$ are shown below (there are 33).

000000	000001	000010	000011	000100
000101	000111	001000	001001	001010
001011	001111	010000	010001	010010
010011	010100	010101	010111	011111
100000	100001	100010	100011	100100
100101	100111	101000	101001	101010
101011	101111	111111		

The command for generating these numbers is:

```
autogen 110not.aut 6 0 0 1 2
```

where `110not.aut` is the automaton file consisting of the following lines:

```
3
0 00 11
1 00 12
2 12
```

Problem 13. Using three different symbols, how many strings of length n are there where no two adjacent symbols are equal?

Solution. This can be solved using simple combinatorial reasoning. There are three choices for the first symbol and two choices for each subsequent symbol.

The total number of strings of length n is then $a(n) = 3 \cdot 2^{n-1}$. We will solve this simple problem using automaton and matrix methods that will be useful in more difficult problems.

Let the three symbols be $1, 2, 3$, then the automaton for generating or recognizing such strings is shown below.

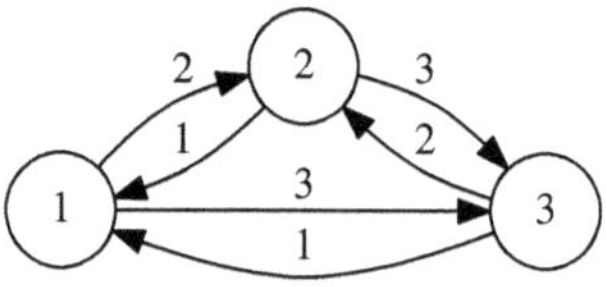

We start in the state corresponding to the first symbol and then make $n-1$ transitions for the rest of the symbols. The number of walks of length n between state i and state j of the automaton is equal to the (i, j) element of the nth power of the automaton's adjacency matrix. The adjacency matrix is

$$\mathbf{A} = \begin{pmatrix} 0 & 1 & 1 \\ 1 & 0 & 1 \\ 1 & 1 & 0 \end{pmatrix} \tag{69}$$

The nth power of the adjacency matrix is equal to the coefficient of z^n in the power series expansion of $(\mathbf{I} - z\mathbf{A})^{-1}$

$$(\mathbf{I} - z\mathbf{A})^{-1} = \mathbf{I} + \mathbf{A}z + \mathbf{A}^2 z^2 + \mathbf{A}^3 z^3 + \dots \tag{70}$$

Calculating the inverse, we get

$$(\mathbf{I}-z\mathbf{A})^{-1} = \frac{1}{(1+z)(1-2z)} \begin{pmatrix} 1-z & z & z \\ z & 1-z & z \\ z & z & 1-z \end{pmatrix} \tag{71}$$

Since a string can start and end at any state, the generating function is found by summing all the elements of the matrix and multiplying the result by z. The multiplication by z represents the first symbol that puts us in the initial state. The reading or generation of the first symbol can be made explicit by putting a start state in the automaton with a transition to states $1, 2, 3$. We have chosen not to do this since it would increase the dimension of the adjacency matrix. The generating function is

$$G(z) = \frac{3z}{1-2z} \tag{72}$$

The Taylor series expansion is

$$3(z + 2z^2 + 4z^3 + 8z^4 + 16z^5 + \ldots) \tag{73}$$

From which we get that the number of strings of length n is $a(n) = 3 \cdot 2^{n-1}$. The integer sequence is A007283 in the OEIS.

All the strings for $n = 5$ are shown below (there are

48).

12121	12123	12131	12132	12312	12313
12321	12323	13121	13123	13131	13132
13212	13213	13231	13232	21212	21213
21231	21232	21312	21313	21321	21323
23121	23123	23131	23132	23212	23213
23231	23232	31212	31213	31231	31232
31312	31313	31321	31323	32121	32123
32131	32132	32312	32313	32321	32323

The command for generating these strings is:

```
autogen 3sym2adjnoteq.aut 5 S 1 2 3
```

where `3sym2adjnoteq.aut` is the automaton file consisting of the following lines:

```
4
S 11 22 33
1 22 33
2 11 33
3 11 22
```

Problem 14. How many strings of length n are there using the symbols $1, 2, 3$ where adjacent symbols never differ by more than 1. In other words, the symbol pairs 13 and 31 never occur.

Solution. The automaton for generating or recognizing such strings is shown below.

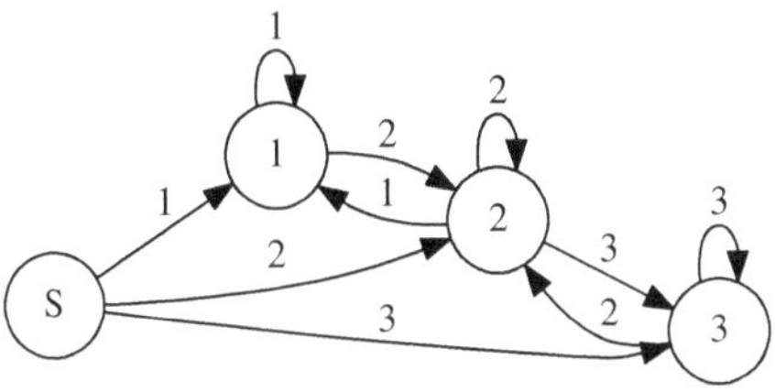

The automaton's adjacency matrix is

$$\mathbf{A} = \begin{pmatrix} 0 & 1 & 1 & 1 \\ 0 & 1 & 1 & 0 \\ 0 & 1 & 1 & 1 \\ 0 & 0 & 1 & 1 \end{pmatrix} \tag{74}$$

The start state is S, and we can end in any of the states, so the generating function is found by summing the first row of $(\mathbf{I} - z\mathbf{A})^{-1}$, which is

$$G(z) = \frac{1+z}{1-2z-z^2} \tag{75}$$

The Taylor series expansion is

$$G(z) = 1+3z+7z^2+17z^3+41z^4+99z^5+239z^6+577z^7 \ldots \tag{76}$$

From which we get that the number of strings of length n is

$$a(n) = \frac{(1-\sqrt{2})^{n+1} + (1+\sqrt{2})^{n+1}}{2} \tag{77}$$

The integer sequence is A001333 in the OEIS.

All the strings for $n = 4$ are shown below (there are 41).

1111	1112	1121	1122	1123	1211	1212
1221	1222	1223	1232	1233	2111	2112
2121	2122	2123	2211	2212	2221	2222
2223	2232	2233	2321	2322	2323	2332
2333	3211	3212	3221	3222	3223	3232
3233	3321	3322	3323	3332	3333	

The command for generating these strings is:

```
autogen 3sym2adjnotdiffer2.aut 4 S 1 2 3
```

where `3sym2adjnotdiffer2.aut` is the automaton file consisting of the following lines:

```
4
S 11 22 33
1 11 22
2 11 22 33
3 22 33
```

Problem 15. How many strings of length n are there using the symbols $1, 2, 3$ where adjacent symbols never differ by more than 1, and are never equal.

Solution. The automaton for generating or recognizing such strings is shown below.

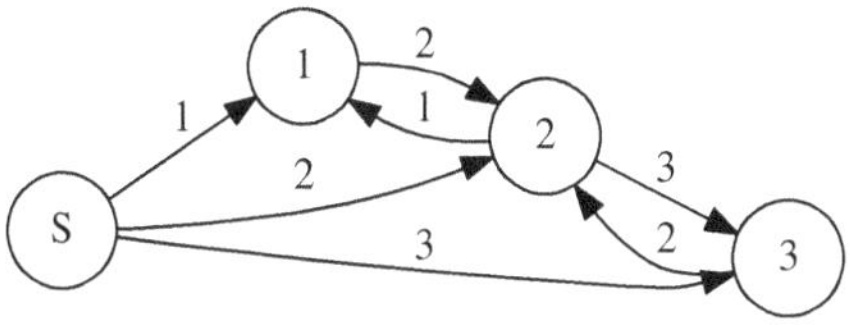

The automaton's adjacency matrix is

$$\mathbf{A} = \begin{pmatrix} 0 & 1 & 1 & 1 \\ 0 & 0 & 1 & 0 \\ 0 & 1 & 0 & 1 \\ 0 & 0 & 1 & 0 \end{pmatrix} \tag{78}$$

The start state is S, and we can end in any of the states, so the generating function is found by summing the first row of $(\mathbf{I} - z\mathbf{A})^{-1}$, which is

$$G(z) = \frac{1 + 3z + 2z^2}{1 - 2z^2} \tag{79}$$

The Taylor series expansion is

$$G(z) = 1 + 3z + 4z^2 + 6z^3 + 8z^4 + 12z^5 + 16z^6 + 24z^7 \ldots \tag{80}$$

From which we get that the number of strings of length n is

$$a(n) = \begin{cases} 1, & n = 0 \\ 2^{(n+2)/2}, & n = 2, 4, 6 \ldots \\ 3 \cdot 2^{(n-1)/2}, & n = 1, 3, 5 \ldots \end{cases} \tag{81}$$

The integer sequence is A063759 in the OEIS.

All the strings for $n = 7$ are shown below (there are 24).

1212121	1212123	1212321	1212323	1232121
1232123	1232321	1232323	2121212	2121232
2123212	2123232	2321212	2321232	2323212
2323232	3212121	3212123	3212321	3212323
3232121	3232123	3232321	3232323	

The command for generating these strings is:

```
autogen 3sym2adjnotdiffer2noteq.aut 7 S 1 2 3
```

where `3sym2adjnotdiffer2noteq.aut` is the automaton file consisting of the following lines:

```
4
S 11 22 33
1 22
2 11 33
3 22
```

Problem 16. Strings are to be composed of symbols from the list $[1, 2, 3, 4]$. Adjacent symbols in the string must be adjacent in the list, which is circular, so that 1 and 4 are considered adjacent. How many such strings of length n are there?

Solution. This can be solved using simple combinatorial reasoning. There are four choices for the first symbol and two choices for each subsequent symbol. The total number of strings of length n is then $a(n) = 4 \cdot 2^{n-1} = 2^{n+1}$ for $n > 0$. Now we solve the problem using the automaton.

The automaton for generating or recognizing such strings is shown below.

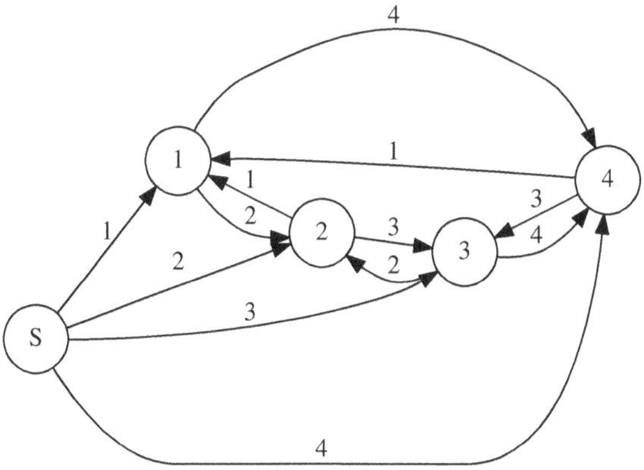

The automaton's adjacency matrix is

$$\mathbf{A} = \begin{pmatrix} 0 & 1 & 1 & 1 & 1 \\ 0 & 0 & 1 & 0 & 1 \\ 0 & 1 & 0 & 1 & 0 \\ 0 & 0 & 1 & 0 & 1 \\ 0 & 1 & 0 & 1 & 0 \end{pmatrix} \qquad (82)$$

The start state is S, and we can end in any of the states, so the generating function is found by summing the first row of $(\mathbf{I} - z\mathbf{A})^{-1}$, which is

$$G(z) = \frac{1+2z}{1-2z} \tag{83}$$

The Taylor series expansion is

$$G(z) = 1+4z+8z^2+16z^3+32z^4+64z^5+128z^6+256z^7 \ldots \tag{84}$$

This matches our combinatorially derived solution, except for the first term (1), which only indicates there is one such string of zero length. The integer sequence is A151821 in the OEIS.

All the strings for $n = 4$ are shown below (there are 32).

1212	1214	1232	1234	1412	1414	1432
1434	2121	2123	2141	2143	2321	2323
2341	2343	3212	3214	3232	3234	3412
3414	3432	3434	4121	4123	4141	4143
4321	4323	4341	4343			

The command for generating these strings is:

```
autogen 4sym2adjlistcirc.aut 4 S 1 2 3 4
```

where `4sym2adjlistcirc.aut` is the automaton file consisting of the following lines:

```
5
S 11 22 33 44
1 22 44
2 11 33
3 22 44
4 11 33
```

Problem 17. Repeat the previous problem, except the even numbers (2, 4) are now allowed to repeat.

Solution. The states are no longer homogeneous, so a combinatorial solution is not trivial. The automaton for generating or recognizing such strings is shown below.

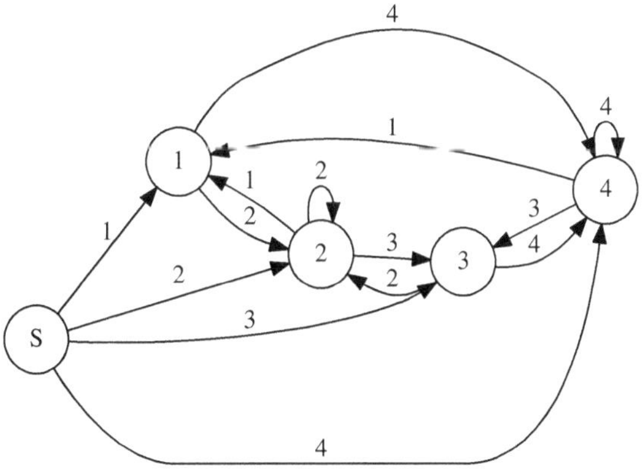

The automaton's adjacency matrix is

$$\mathbf{A} = \begin{pmatrix} 0 & 1 & 1 & 1 & 1 \\ 0 & 0 & 1 & 0 & 1 \\ 0 & 1 & 1 & 1 & 0 \\ 0 & 0 & 1 & 0 & 1 \\ 0 & 1 & 0 & 1 & 1 \end{pmatrix} \tag{85}$$

The start state is S, and we can end in any of the states, so the generating function is found by summing the first row of $(\mathbf{I} - z\mathbf{A})^{-1}$, which is

$$G(z) = \frac{1 + 3z + 2z^2}{1 - z - 4z^2} \tag{86}$$

The Taylor series expansion is

$$G(z) = 1{+}4z{+}10z^2{+}26z^3{+}66z^4{+}170z^5{+}434z^6{+}1114z^7 \ldots \tag{87}$$

The values for strings of length 0 through 12 are

n	0	1	2	3	4	5	6	7
$a(n)$	1	4	10	26	66	170	434	1114

n	8	9	10	11	12
$a(n)$	2850	7306	18706	47930	122754

The integer sequence is A277236 in the OEIS.

For $n \geq 3$, the integer sequence obeys the simple recurrence

$$a(n) = a(n-1) + 4a(n-2) \tag{88}$$

with the initial values $a(1) = 4$ and $a(2) = 10$.

All the strings for $n = 4$ are shown below (there are 66).

1212	1214	1221	1222	1223	1232	1234
1412	1414	1432	1434	1441	1443	1444
2121	2122	2123	2141	2143	2144	2212
2214	2221	2222	2223	2232	2234	2321
2322	2323	2341	2343	2344	3212	3214
3221	3222	3223	3232	3234	3412	3414
3432	3434	3441	3443	3444	4121	4122
4123	4141	4143	4144	4321	4322	4323
4341	4343	4344	4412	4414	4432	4434
4441	4443	4444				

The command for generating these strings is:

```
autogen 4sym2adjlistcirceven.aut 4 S 1 2 3 4
```

where `4sym2adjlistcirceven.aut` is the automaton file consisting of the following lines:

```
5
S 11 22 33 44
1 22 44
2 11 22 33
3 22 44
4 11 33 44
```

Problem 18. Repeat the previous problem, except we add two more symbols so that strings are composed of

$[1, 2, 3, 4, 5, 6]$ with even numbers (2, 4, 6) allowed to repeat.

Solution. The automaton for generating or recognizing such strings is shown below.

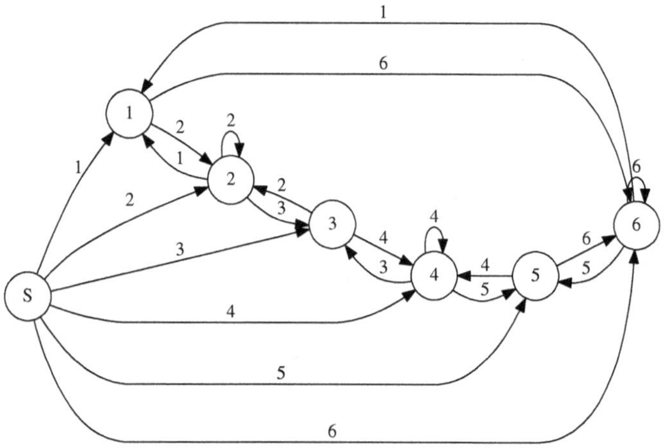

The automaton's adjacency matrix is

$$\mathbf{A} = \begin{pmatrix} 0 & 1 & 1 & 1 & 1 & 1 & 1 \\ 0 & 0 & 1 & 0 & 0 & 0 & 1 \\ 0 & 1 & 1 & 1 & 0 & 0 & 0 \\ 0 & 0 & 1 & 0 & 1 & 0 & 0 \\ 0 & 0 & 0 & 1 & 1 & 1 & 0 \\ 0 & 0 & 0 & 0 & 1 & 0 & 1 \\ 0 & 1 & 0 & 0 & 0 & 1 & 1 \end{pmatrix} \tag{89}$$

The start state is S, and we can end in any of the states, so the generating function is found by summing the first row of $(\mathbf{I} - z\mathbf{A})^{-1}$, which is

$$G(z) = \frac{1 + 5z + 5z^2}{1 - z - 4z^2} \tag{90}$$

The Taylor series expansion is

$$G(z) = 1+6z+15z^2+39z^3+99z^4+255z^5+651z^6+1671z^7 \ldots \tag{91}$$

The values for strings of length 0 through 12 are

n	0	1	2	3	4	5	6	7
$a(n)$	1	6	15	39	99	255	651	1671

n	8	9	10	11	12
$a(n)$	4275	10959	28059	71895	184131

The integer sequence is A277237 in the OEIS.

For $n \geq 3$, the integer sequence obeys the simple recurrence

$$a(n) = a(n-1) + 4a(n-2) \tag{92}$$

with the initial values $a(1) = 6$ and $a(2) = 15$.

All the strings for $n = 4$ are shown below (there are

99).

1212	1216	1221	1222	1223	1232	1234	1612
1616	1654	1656	1661	1665	1666	2121	2122
2123	2161	2165	2166	2212	2216	2221	2222
2223	2232	2234	2321	2322	2323	2341	2343
2344	3212	3216	3221	3222	3223	3232	3234
3412	3416	3432	3434	3441	3443	3444	4121
4122	4123	4161	4165	4166	4321	4322	4323
4341	4343	4344	4412	4416	4432	4434	4441
4443	4444	5412	5416	5432	5434	5441	5443
5444	5612	5616	5654	5656	5661	5665	5666
6121	6122	6123	6161	6165	6166	6541	6543
6544	6561	6565	6566	6612	6616	6654	6656
6661	6665	6666					

The command for generating these strings is:

```
autogen 6sym2adjlistcirceven.aut 4 S 1 2 3 4 5 6
```

where `6sym2adjlistcirceven.aut` is the automaton file consisting of the following lines:

```
7
S 11 22 33 44 55 66
1 22 66
2 11 22 33
3 22 44
4 11 33 44
5 44 66
6 11 55 66
```

Problem 19. Repeat the previous problem, except we have only two symbols so that strings are composed of [1, 2] with only 2 allowed to repeat.

Solution. The automaton for generating or recognizing such strings is shown below.

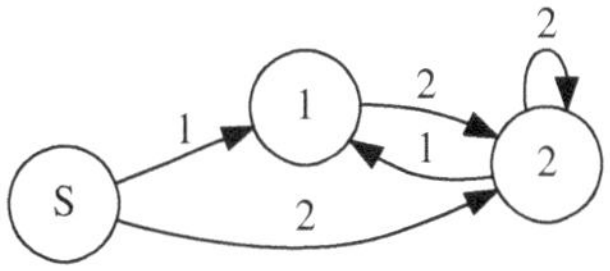

The automaton's adjacency matrix is

$$\mathbf{A} = \begin{pmatrix} 0 & 1 & 1 \\ 0 & 0 & 1 \\ 0 & 1 & 1 \end{pmatrix} \tag{93}$$

The start state is S, and we can end in any of the states, so the generating function is found by summing the first row of $(\mathbf{I} - z\mathbf{A})^{-1}$, which is

$$G(z) = \frac{1+z}{1-z-z^2} \tag{94}$$

The Taylor series expansion is

$$G(z) = 1+2z+3z^2+5z^3+8z^4+13z^5+21z^6+34z^7+55z^8 \dots$$

(95)

The integer sequence is A000045 in the OEIS. The numbers are the well known Fibonacci numbers.

Problem 20. Find the cycle index polynomial for strings of length n with respect to reversal symmetry, i.e. two strings are considered identical if one is the reverse of the other.

Solution. There are two symmetry operations in this problem, the identity and reversal. Let the string positions be numbered consecutively $1, 2, \ldots, n$.

For $n = 3$, the cycle structure for the identity is $(1)(2)(3)$, and for reversal, it is $(1\ 3)(2)$. The cycle index polynomial is then

$$p(z) = \frac{1}{2}(z_1^3 + z_1 z_2)$$

For $n = 4$, the identity and reversal have cycle structure $(1)(2)(3)(4)$ and $(1\ 4)(2\ 3)$, respectively. The cycle index polynomial for $n = 4$ is

$$p(z) = \frac{1}{2}(z_1^4 + z_2^2)$$

The pattern for odd and even values of n is clear. The cycle index polynomial for n is

$$p(z) = \frac{1}{2}\begin{cases} z_1^n + z_2^{n/2}, & n = 0, 2, 4, \ldots \\ z_1^n + z_1 z_2^{(n-1)/2}, & n = 1, 3, 5, \ldots \end{cases} \tag{96}$$

Problem 21. Two strings of length n are considered identical if one is the reverse of the other. How many unique strings of length n are there using m kinds of letters? Calculate the number of strings for $m = 2$ and $n = 1, \ldots, 10$. Find all the strings for $m = 2$ and $n = 6$.

Solution. The cycle index polynomial for this problem was found in the previous problem. To get the number of strings, make the substitution $z_i \to m$ in equation 96. The number of strings as a function of n and m is then

$$p(n,m) = \frac{1}{2}\begin{cases} m^n + m^{n/2}, & n = 0, 2, 4, \ldots \\ m^n + m^{(n+1)/2}, & n = 1, 3, 5, \ldots \end{cases} \tag{97}$$

For $m = 2$, the values for strings of length 1 through 10 are

n	1	2	3	4	5	6	7	8	9	10
$p(n,2)$	2	3	6	10	20	36	72	136	272	528

The integer sequence is A005418 in the OEIS.

All the strings for $m = 2$ and $n = 6$ are shown below (there are 36).

000000	000001	000010	000011	000100	000101
000110	000111	001001	001010	001011	001100
001101	001110	001111	010001	010010	010011
010101	010110	010111	011001	011011	011101
011110	011111	100001	100011	100101	100111
101011	101101	101111	110011	110111	111111

The command for generating these strings is:

```
autogen binary.aut 6 0 0 1 | revfilt 6
```

where `binary.aut` is the automaton file consisting of the following lines:

```
2
0 00 11
1 00 11
```

Problem 22. The square tiles in a linear array of n tiles are colored black and white. Two colorings are considered the same if one is the reverse of the other. How many ways can the tiles be colored so that there are exactly two black tiles. Find all the tilings for $n = 10$.

Solution. The cycle index polynomial for this problem is given by equation 96. To find the number of colorings, we could substitute $z_i \to b^i + w^i$ into equation 96 and then find the coefficient of b^2w^{n-2}, but we can simplify things a bit by setting $w = 1$ and using the substitution $z_i \to b^i + 1$. This gives us

$$p_n(b) = \frac{1}{2}\begin{cases} (b+1)^n + (b^2+1)^{n/2}, & n = 0, 2, 4, \ldots \\ (b+1)^n + (b+1)(b^2+1)^{(n-1)/2}, & n = 1, 3, 5, \ldots \end{cases} \tag{98}$$

Now all we have to do is find the coefficient of b^2 in this equation. For even n, the coefficient of b^2 in $(b+1)^n$ is $\binom{n}{2}$ and the coefficient of b^2 in $(b^2+1)^{n/2}$ is just $n/2$. So for even n, the answer is

$$\frac{\binom{n}{2} + \frac{n}{2}}{2}$$

Likewise, for odd n, the answer is

$$\frac{\binom{n}{2} + \frac{n-1}{2}}{2}$$

The values for strings of length 1 through 10 are

n	1	2	3	4	5	6	7	8	9	10
$f(n)$	0	1	2	4	6	9	12	16	20	25

The integer sequence is A002620 in the OEIS.

All the tilings for $n = 10$ are shown below (there are 25).

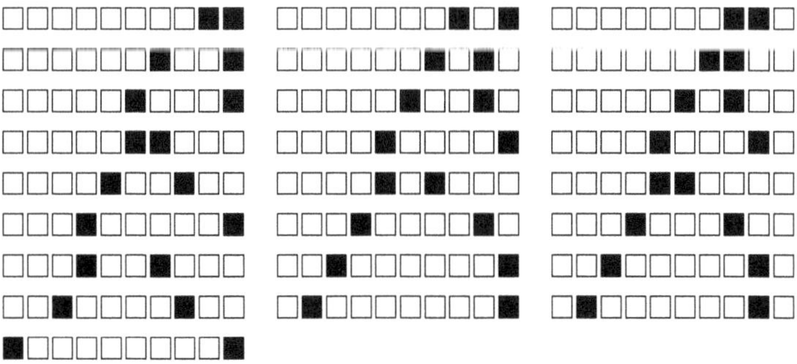

The command for generating these tilings is:

```
autogen 2blacktiles.aut 10 0 2 | revfilt 10
```

where `2blacktiles.aut` is the automaton file consisting of the following lines:

```
3
0 00 11
1 01 12
2 02
```

Problem 23. Generalize the previous problem to find the number of ways the tiles can be colored so that there are k black tiles.

Solution. The solution is given by the coefficient of b^k in equation 98. The four cases have to be considered where n is even and odd, and k is even and odd. When n is even, the coefficient of b^k in equation 98 is

$$f_k(n) = \frac{1}{2}\begin{cases} \binom{n}{k} + \binom{n/2}{k/2}, & k = 0, 2, 4, \ldots \\ \binom{n}{k}, & k = 1, 3, 5, \ldots \end{cases} \tag{99}$$

When n is odd, the coefficient of b^k in equation 98 is

$$f_k(n) = \frac{1}{2}\begin{cases} \binom{n}{k} + \binom{(n-1)/2}{k/2}, & k = 0, 2, 4, \ldots \\ \binom{n}{k} + \binom{(n-1)/2}{(k-1)/2}, & k = 1, 3, 5, \ldots \end{cases} \tag{100}$$

The values for $n = 0, 1, \ldots 10$ and $k = 0, 1, \ldots n$ are shown in the table below. The table can be found as integer sequence A034851 in the OEIS. It is a well known sequence called Losanitsch's triangle.

k / n	0	1	2	3	4	5	6	7	8	9	10
0	1										
1	1	1									
2	1	1	1								
3	1	2	2	1							
4	1	2	4	2	1						
5	1	3	6	6	3	1					
6	1	3	9	10	9	3	1				
7	1	4	12	19	19	12	4	1			
8	1	4	16	28	38	28	16	4	1		
9	1	5	20	44	66	66	44	20	5	1	
10	1	5	25	60	110	126	110	60	25	5	1

Problem 24. Find the number of binary strings of length n that do not contain two or more consecutive 1's, and where two strings are considered equivalent if one is the reverse of the other. Find all the strings for $n = 8$.

Solution. The way to solve this problem is to use Burnside's Theorem which says that the number of equivalence classes induced on the strings by a group of permutations G is given by

$$\frac{1}{|G|} = \sum_{g \in G} \psi(g)$$

where $|G|$ is the size of the group, g is one of the permutations in G, and $\psi(g)$ is the number of strings that are invariant under the permutation g.

For this problem there are only two permutations in the group G. The identity, which we will call g_0, and the permutation that reverses the string, which we will call g_1. The number of equivalence classes, or in other words, the number of unique strings under reversal symmetry is given by

$$\frac{1}{2}\left(\psi(g_0) + \psi(g_1)\right)$$

For now, let's call the number of strings without the reversal equivalence F_n. These are the number of binary strings that do not contain two or more consecutive 1's which we will call Fibonacci strings. All the strings are invariant with respect to the identity so that $\psi(g_0) = F_n$. To find the number of Fibonacci strings invariant with respect to reversal, we need to look at the case where n is even and odd, separately.

For $n = 2m$, a string invariant with respect to reversal will look like $b_1b_2 \dots b_m b_m \dots b_2 b_1$ where b_i is a binary digit equal to 0 or 1. The string $b_1 b_2 \dots b_m$ is a Fibonacci string where $b_m = 0$. The number of such strings is F_{m-1}. So for $n = 2m$ we have $\psi(g_1) = F_{m-1} = F_{n/2-1}$ and the number of unique strings is

$$\frac{1}{2}\left(F_n + F_{n/2-1}\right)$$

For $n = 2m + 1$, a string invariant with respect to reversal will look like $b_1 b_2 \dots b_m c b_m \dots b_2 b_1$ where c and b_i

are binary digits equal to 0 or 1. The string $b_1b_2\dots b_mc$ is a Fibonacci string of length $m+1$. The number of such strings is F_{m+1}. So for $n = 2m+1$ we have $\psi(g_1) = F_{m+1} = F_{(n+1)/2}$ and the number of unique strings is

$$\frac{1}{2}\left(F_n + F_{(n+1)/2}\right)$$

The only thing we need to do now is find a way to calculate the numbers F_i. An automaton for generating Fibonacci strings is shown in the figure below.

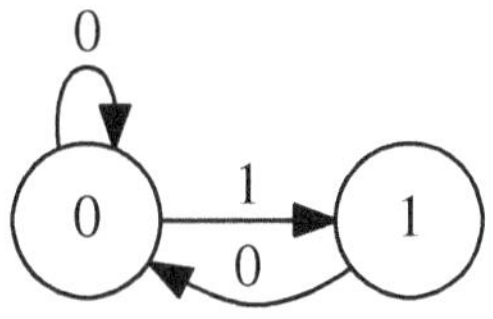

The adjacency matrix for the automaton is

$$\mathbf{A} = \begin{pmatrix} 1 & 1 \\ 1 & 0 \end{pmatrix} \tag{101}$$

A Fibonacci string is generated by starting in state 0 and ending in state 0 or 1, so we get the generating function for the number of these strings by calculating $(\mathbf{I} - z\mathbf{A})^{-1}$ and summing the first row of the result.

$$(\mathbf{I} - z\mathbf{A})^{-1} = \frac{1}{1 - z - z^2} \begin{pmatrix} 1 & z \\ z & 1 - z \end{pmatrix}$$

The generating function is then

$$G(z) = \frac{1 + z}{1 - z - z^2} \tag{102}$$

The Taylor series expansion is

$$G(z) = 1+2z+3z^2+5z^3+8z^4+13z^5+21z^6+34z^7+55z^8 \ldots \tag{103}$$

The integer sequence is A000045 in the OEIS. The numbers are the well known Fibonacci numbers.

The recurrence equation for these numbers is $F_n = F_{n-1} + F_{n-2}$ with starting values $F_0 = 1$ and $F_1 = 2$. Using these values in the above equations, the number of Fibonacci strings under reversal equivalence is

n	0	1	2	3	4	5	6	7	8	9	10
$a(n)$	1	2	2	4	5	9	12	21	30	51	76

The integer sequence is A001224 in the OEIS.

All the strings for $n = 8$ are shown below, $\blacksquare = 1$ and $\square = 0$ (there are 30).

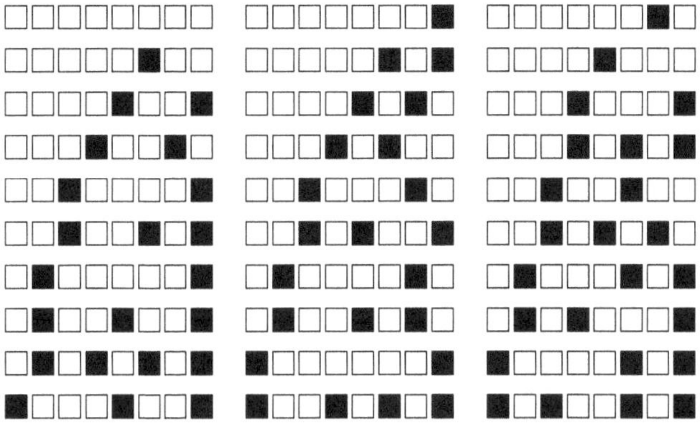

The command for generating these strings is:

```
autogen fibo.aut 8 0 0 1 | revfilt 8
```

where `fibo.aut` is the automaton file consisting of the following lines:

```
2
0 00 11
1 00
```

Problem 25. Find the number of binary strings of length n that do not contain three or more consecutive 1's, and where two strings are considered equivalent if one is the reverse of the other. Find all the strings for $n = 7$.

Solution. This is just like the previous problem where we used Burnside's Theorem. We have two group elements, the identity g_0, and the permutation that reverses the string g_1. The number of unique strings is then given by

$$a(n) = \frac{1}{2}\left(\psi(g_0) + \psi(g_1)\right)$$

We'll start by finding the number of strings without the reversal equivalence. An automaton for generating these strings is shown below.

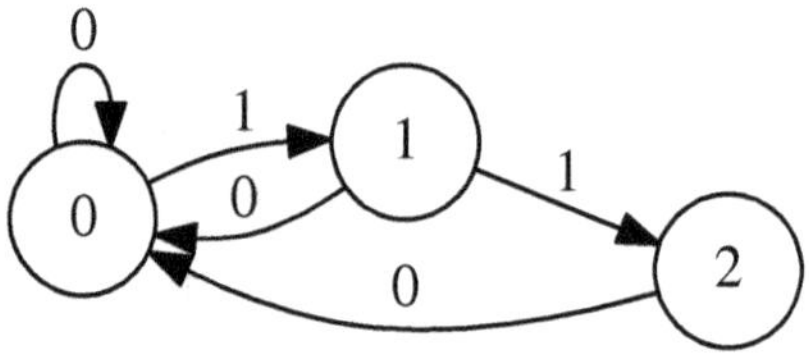

Starting in state 0 and taking all possible paths of length n that end in state 0, 1 or 2, will generate all binary strings of length n that do not contain three or more consecutive 1's. We will call these tribonacci strings. The adjacency matrix for the automaton is

$$\mathbf{A} = \begin{pmatrix} 1 & 1 & 0 \\ 1 & 0 & 1 \\ 1 & 0 & 0 \end{pmatrix} \tag{104}$$

To find the generating function, we calculate

$$(\mathbf{I}-z\mathbf{A})^{-1} = \frac{1}{1-z-z^2-z^3}\begin{pmatrix} 1 & z & z^2 \\ z+z^2 & 1-z & z-z^2 \\ z & z^2 & 1-z-z^2 \end{pmatrix}$$

The generating function is the sum of the first row

$$G(z) = \frac{1+z+z^2}{1-z-z^2-z^3} \tag{105}$$

The Taylor series expansion is

$$\begin{aligned} G(z) &= 1+2z+4z^2+7z^3+13z^4+24z^5+44z^6+ \\ & \quad 81z^7+149z^8\ldots \end{aligned} \tag{106}$$

The integer sequence is A000073 in the OEIS. These are the well known tribonacci numbers. They obey a simple recurrence. Let T_n be the nth number in the sequence.[6] then $T_n = T_{n-1} + T_{n-2} + T_{n-3}$ with initial values $T_0 = 1$, $T_1 = 2$, and $T_2 = 4$.

All the strings are invariant with respect to the identity so $\psi(g_0) = T_n$. To find the number invariant with respect to reversal, we need to look at the case of odd and even length separately.

For $n = 2m$, the invariant strings look like
$b_1b_2\ldots b_mb_m\ldots b_2b_1$
where $b_1b_2\ldots b_m$ is a tribonacci string with $b_{m-1}b_m \neq 11$, i.e. the string cannot end with two 1's. There are

[6]This is the number of tribonacci strings of length n

two possibilities. If $b_m = 0$, then the string $b_1b_2 \ldots b_{m-1}$ can be any tribonacci string and there are T_{m-1} of them. If $b_{m-1}b_m = 01$, then the string $b_1b_2 \ldots b_{m-2}$ can be any tribonacci string and there are T_{m-2} of them. The number of invariant strings is then

$$\psi(g_1) = T_{m-1} + T_{m-2} = T_{n/2-1} + T_{n/2-2}$$

For $n = 2m + 1$, the invariant strings look like $b_1b_2 \ldots b_m c b_m \ldots b_2 b_1$.
If $c = 0$ then $b_1b_2 \ldots b_m$ can be any tribonacci string, and there are T_m of them. If $c = 1$ then we must have $b_m = 0$, and $b_1b_2 \ldots b_{m-1}$ can be any tribonacci string. The number of invariant strings is then

$$\psi(g_1) = T_m + T_{m-1} = T_{(n-1)/2} + T_{(n-3)/2}$$

The number of tribonacci strings, taking into account reversal equivalence, is then:

$$a(n) = \frac{1}{2}\begin{cases} T_n + T_{n/2-1} + T_{n/2-2}, & n = \text{even} \\ T_n + T_{(n-1)/2} + T_{(n-3)/2}, & n = \text{odd} \end{cases} \tag{107}$$

The first few terms in the sequence are

n	0	1	2	3	4	5	6	7	8	9	10
$a(n)$	1	2	3	5	8	15	25	46	80	147	262

The integer sequence is A151518 in the OEIS.

All the tribonacci strings for $n = 7$ are shown below, ■ = 1 and □ = 0 (there are 46).

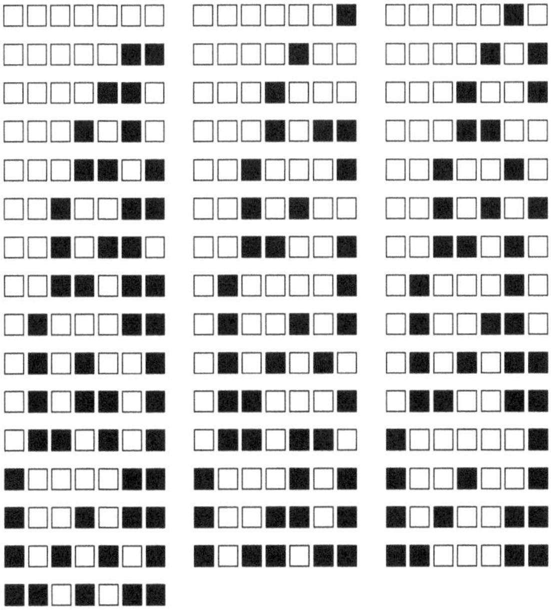

The command for generating the example above is:

```
autogen tribo.aut 7 0 0 1 2 | revfilt 7
```

where `tribo.aut` defines the automaton, consisting of the following lines:

```
3
0 00 11
1 00 12
2 00
```

Problem 26. Take a binary string and connect the end to the beginning to get a circular binary string. Find the number of circular binary strings of length n that do not contain two or more consecutive 1's. Find all strings of length $n = 8$.

Solution. We have looked at strings that do not contain two or more consecutive ones in previous problems, where they were called Fibonacci strings. In this problem we want to count the number of circular Fibonacci strings. These are all the regular Fibonacci strings that do not begin and end with a 1. The automaton for generating these strings is shown below.

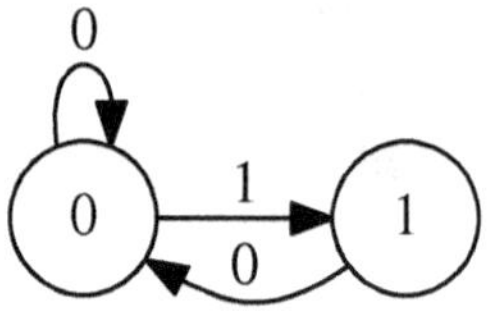

The adjacency matrix for the automaton is

$$\mathbf{A} = \begin{pmatrix} 1 & 1 \\ 1 & 0 \end{pmatrix} \tag{108}$$

This is the same automaton used to generate regular Fibonacci strings. A circular string is generated by starting and ending in state 0, or by starting and ending in state 1. So we get the generating function for the number of circular strings by calculating $(\mathbf{I}-z\mathbf{A})^{-1}$ and summing the diagonal elements, which is also called the trace, and is denoted by Tr $(\mathbf{I}-z\mathbf{A})^{-1}$.

$$(\mathbf{I}-z\mathbf{A})^{-1} = \frac{1}{1-z-z^2}\begin{pmatrix} 1 & z \\ z & 1-z \end{pmatrix} \tag{109}$$

The generating function is then

$$G(z) = \frac{2-z}{1-z-z^2} \tag{110}$$

The Taylor series expansion is

$$G(z) = 2+z+3z^2+4z^3+7z^4+11z^5+18z^6+29z^7+47z^8\ldots \tag{111}$$

The integer sequence is A000032 in the OEIS. The numbers are the well known Lucas numbers. They obey the same recurrence as the Fibonacci numbers, but with different starting values. Let L_n be the number of circular Fibonacci strings of length n, then $L_n = L_{n-1} + L_{n-2}$ with $L_0 = 2$ and $L_1 = 1$. The following table lists the first ten values.

n	0	1	2	3	4	5	6	7	8	9	10
$L(n)$	2	1	3	4	7	11	18	29	47	76	123

All the strings for $n = 8$ are shown below, ■ = 1 and □ = 0 (there are 47).

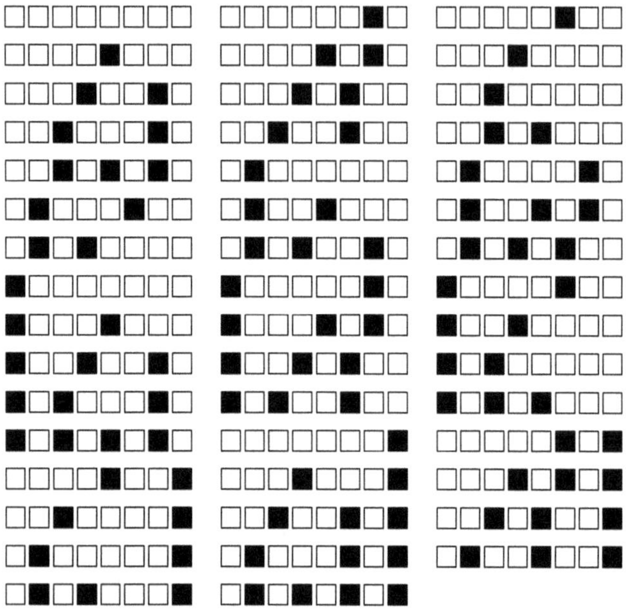

The example above is generated by combining the results of the following two commands:

```
autogen fibo.aut 8 0 0
autogen fibo.aut 8 1 1
```

where `fibo.aut` defines the automaton, consisting of the following lines:

```
2
0 00 11
1 00
```

Problem 27. Similar to the previous problem, we want to find the number of circular binary strings of length n that do not contain three or more consecutive 1's. Find all strings of length $n = 7$.

Solution. We have looked at strings that do not contain three or more consecutive ones in previous problems, where they were called tribonacci strings. In this problem we want to count the number of circular tribonacci strings. The automaton for generating these strings is shown below.

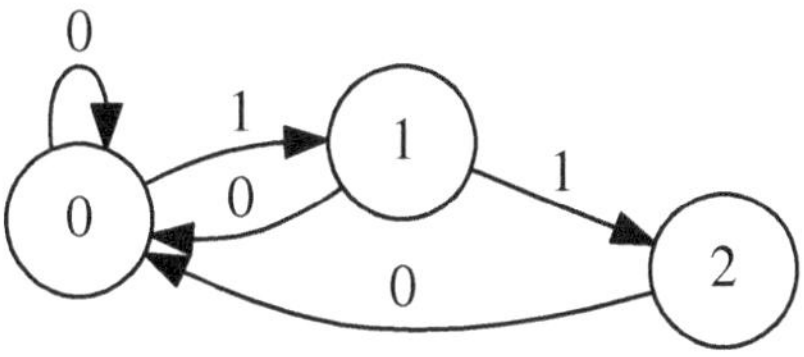

The adjacency matrix for the automaton is

$$\mathbf{A} = \begin{pmatrix} 1 & 1 & 0 \\ 1 & 0 & 1 \\ 1 & 0 & 0 \end{pmatrix} \tag{112}$$

This is the same automaton used to generate regular tribonacci strings. As in the previous problem, the generating function for the number of circular strings is equal to the trace (sum of the diagonal elements) of $(\mathbf{I} - z\mathbf{A})^{-1}$.

$$(\mathbf{I}-z\mathbf{A})^{-1} = \frac{1}{1-z-z^2-z^3}\begin{pmatrix} 1 & z & z^2 \\ z+z^2 & 1-z & z-z^2 \\ z & z^2 & 1-z-z^2 \end{pmatrix}$$

The generating function is then

$$G(z) = \frac{3-2z-z^2}{1-z-z^2-z^3} \tag{113}$$

The Taylor series expansion is

$$G(z) = 3+z+3z^2+7z^3+11z^4+21z^5+39z^6+71z^7+131z^8\ldots \tag{114}$$

The integer sequence is A001644 in the OEIS. The numbers obey the same recurrence as the tribonacci numbers, but with different starting values. Let R_n be the number of circular tribonacci strings of length n, then $R_n = R_{n-1} + R_{n-2} + R_{n-3}$ with $R_0 = 3$, $R_1 = 1$ and $R_2 = 3$. The following table lists the first ten values.

n	0	1	2	3	4	5	6	7	8	9	10
$R(n)$	3	1	3	7	11	21	39	71	131	241	443

All the strings for $n = 7$ are shown below, ■ = 1 and □ = 0 (there are 71).

□□□□□□□	□□□□□■□	□□□□■□□	□□□□■■□
□□□■□□□	□□□■□■□	□□□■■□□	□□■□□□□
□□■□□■□	□□■□■□□	□□■□■■□	□□■■□□□
□□■■□■□	□■□□□□□	□■□□□■□	□■□□■□□
□■□□■■□	□■□■□□□	□■□■□■□	□■□■■□□
□■■□□□□	□■■□□■□	□■■□■□□	□■■□■■□
■□□□□□□	■□□□□■□	■□□□■□□	■□□□■■□
■□□■□□□	■□□■□■□	■□□■■□□	■□■□□□□
■□■□□■□	■□■□■□□	■□■□■■□	■□■■□□□
■□■■□■□	■■□□□□□	■■□□□■□	■■□□■□□
■■□□■■□	■■□■□□□	■■□■□■□	■■□■■□□
□□□□□□■	□□□□■□■	□□□■□□■	□□□■■□■
□□■□□□■	□□■□■□■	□□■■□□■	□■□□□□■
□■□□■□■	□■□■□□■	□■□■■□■	□■■□□□■
□■■□■□■	■□□□□□■	■□□□■□■	■□□■□□■
■□□■■□■	■□■□□□■	■□■□■□■	■□■■□□■
□□□□□■■	□□□■□■■	□□■□□■■	□□■■□■■
□■□□□■■	□■□■□■■	□■■□□■■	

The example above is generated by combining the results of the following three commands:

```
autogen tribo.aut 8 0 0
autogen tribo.aut 8 1 1
autogen tribo.aut 8 2 2
```

where `tribo.aut` defines the automaton, consisting of the following lines:

```
3
0 00 11
1 00 12
2 00
```

Problem 28. If you take the circular Fibonacci words of length n (calculated previously) and you equate those words that are circular shifts of each other, then the number of unique words you are left with are the Fibonacci necklaces. Calculate the number of Fibonacci necklaces of length n. Find all necklaces of length $n = 9$.

Solution. This problem must be solved using Burnside's Theorem. The symmetry group is the cyclic group C_n composed of the elements g_i, $i = 0, 1, \ldots n-1$ where g_i is the permutation that rotates everything by i positions. Burnside's Theorem says that the number of unique necklaces is

$$a(n) = \frac{1}{n}\sum_{i=0}^{n-1}\psi(g_i) \tag{115}$$

$\psi(g_i)$ is the number of circular Fibonacci words invariant under a rotation by i positions. A circular Fibonacci word of length n will be invariant under rotation by i positions if it is composed of circular Fibonacci words of length i. Each permutation will be composed of cycles of equal length. The length of the cycles must be a divisor of n. The number of permutations composed of cycles of length d is given by the totient function $\phi(d)$. Using these facts the above equation becomes

$$a(n) = \frac{1}{n}\sum_{d|n} \phi(d) L_{n/d} \tag{116}$$

where L_d are the number of circular Fibonacci words of length d calculated in a previous problem. Evaluating the equation, we get the following numbers.

n	1	2	3	4	5	6	7	8	9	10	11	12
$a(n)$	1	2	2	3	3	5	5	8	10	15	19	31

The integer sequence is A000358 in the OEIS.

All the Fibonacci necklaces for $n = 9$ are shown below, as a nine sided polygon, gray=0 and white=1 (there are 10).

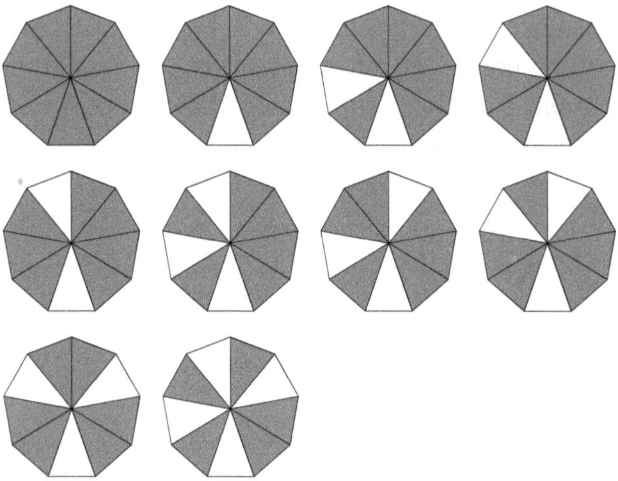

The example above is generated by the following command:

```
kneck 2 9 | autogrep fibo.aut 0 0 1
```

where `fibo.aut` was defined previously.

Problem 29. If you take the circular tribonacci words of length n (calculated previously) and you equate those words that are circular shifts of each other, then the number of unique words you are left with are the tribonacci necklaces. Calculate the number of tribonacci necklaces of length n. Find all necklaces of length $n = 7$.

Solution. This problem must be solved using Burnside's Theorem. The symmetry group is the cyclic group C_n composed of the elements g_i, $i = 0, 1, \ldots n-1$ where g_i is the permutation that rotates everything by i positions. Burnside's Theorem says that the number of unique necklaces is

$$a(n) = \frac{1}{n} \sum_{i=0}^{n-1} \psi(g_i) \tag{117}$$

$\psi(g_i)$ is the number of circular tribonacci words invariant under a rotation by i positions. A circular tribonacci word of length n will be invariant under rotation by i positions if it is composed of circular tribonacci words of length i. Each permutation will be composed of cycles of equal length. The length of the cycles must be a divisor of n. The number of permutations composed of cycles of length d is given by the totient function $\phi(d)$. Using these facts the above equation becomes

$$a(n) = \frac{1}{n} \sum_{d|n} \phi(d) R_{n/d} \tag{118}$$

where R_d are the number of circular tribonacci words of length d calculated in a previous problem. Evaluating the equation, we get the following numbers.

n	1	2	3	4	5	6	7	8	9	10	11	12
$a(n)$	1	2	3	4	5	9	11	19	29	48	75	132

The integer sequence is A093305 in the OEIS.

All the tribonacci necklaces for $n = 7$ are shown below as a seven sided polygon, gray=0 and white=1 (there are 11).

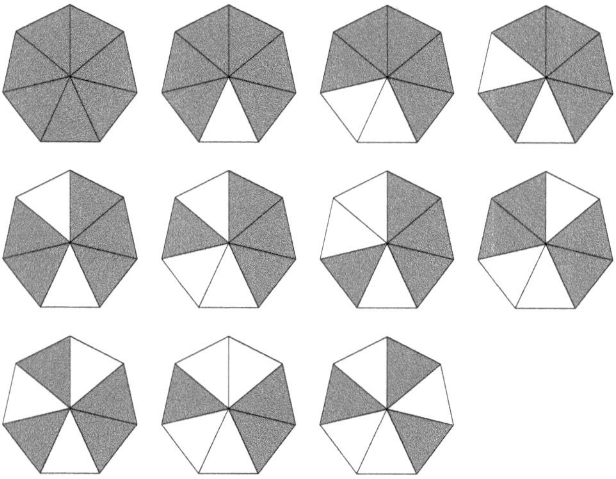

The example above is generated by the following command:

```
kneck 2 7 | autogrep tribo.aut 0 0 1 2
```

where `tribo.aut` was defined previously.

Problem 30. Find the number of binary necklaces of length n with equal number of 1's and 0's.

Solution. We can use equation 38 in the section on Polya and Burnside's Theorem to solve this problem. In this case $f(n/d)$ is equal to the number of binary words of length n/d that have an equal number of 1's and 0's. When n/d is even, the number is $\binom{n/d}{n/2d}$, and when n/d is odd, the number is 0.

$$a(n) = \frac{1}{n}\sum_{d|n}\phi(d)\begin{cases}\binom{n/d}{n/2d}, & n/d = \text{even}\\ 0, & n/d = \text{odd}\end{cases} \tag{119}$$

Evaluating the equation, we get the following numbers, with $a(n) = 0$ when n is odd.

n	2	4	6	8	10	12	14	16	18	20
$a(n)$	1	2	4	10	26	80	246	810	2704	9252

The integer sequence is A003239 in the OEIS.

All the necklaces for $n = 10$ are shown below as a ten sided polygon, gray=1 and white=0 (there are 26).

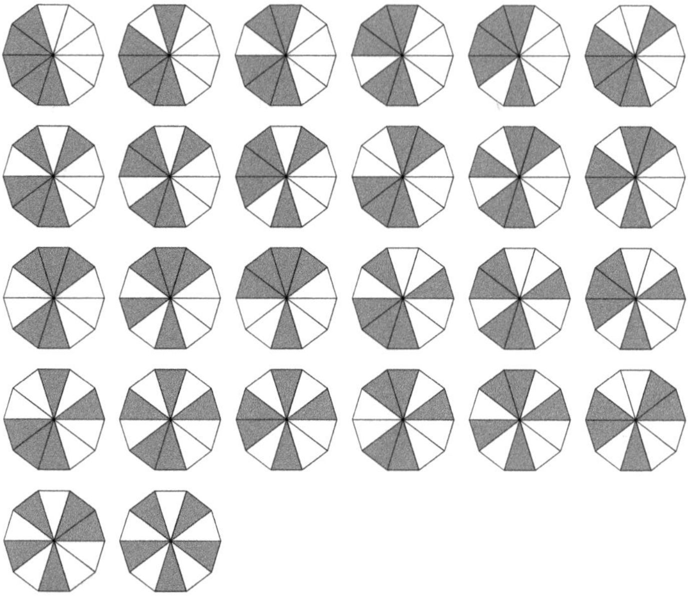

The example above is generated by the following command:

```
kneck 2 10 | filterones.awk numones=5
```

where `filterones.awk` is an awk script that filters binary words based on a specified number of 1's.

Problem 31. Find the number of circular strings of length n composed of the numbers $\{1, 2, 3\}$ where no two adjacent numbers are equal. Find all strings of length $n = 6$.

Solution. The automaton for generating these strings is shown below.

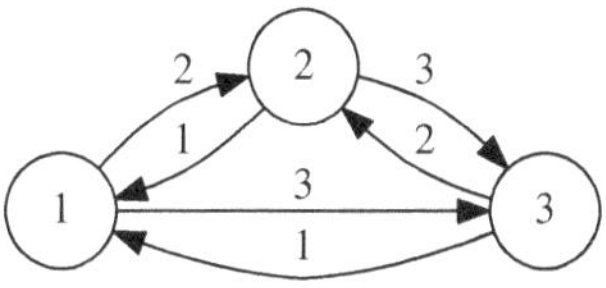

The adjacency matrix for the automaton is

$$\mathbf{A} = \begin{pmatrix} 0 & 1 & 1 \\ 1 & 0 & 1 \\ 1 & 1 & 0 \end{pmatrix} \tag{120}$$

A word of length 2 or more is generated by starting and stopping in state i for $i = 1, 2, 3$. So we get a generating function for these words by taking the trace of $(\mathbf{I} - z\mathbf{A})^{-1}$.

$$(\mathbf{I} - z\mathbf{A})^{-1} = \frac{1}{(1+z)(1-2z)} \begin{pmatrix} 1-z & z & z \\ z & 1-z & z \\ z & z & 1-z \end{pmatrix} \tag{121}$$

The generating function is then

$$G(z) = \frac{3(1-z)}{(1+z)(1-2z)} \tag{122}$$

The Taylor series expansion is

$$G(z) = 3+6z^2+6z^3+18z^4+30z^5+66z^6+126z^7+258z^8 \dots \quad (123)$$

The equation for the nth term is

$$a(n) = \begin{cases} 3, & n = 0 \\ 2^n + 2(-1)^n, & n > 0 \end{cases} \quad (124)$$

This is sequence A092297 in the OEIS.

All the circular strings for $n = 6$ are shown below as tilings, gray=1, white=2 and black=3 (there are 66).

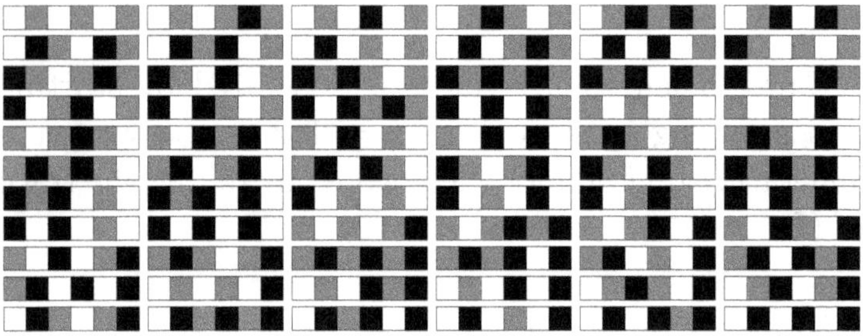

The example above is generated by combining the results of the following commands:

```
autogen 3sym2adjnoteqcirc.aut 6 1 1
autogen 3sym2adjnoteqcirc.aut 6 2 2
autogen 3sym2adjnoteqcirc.aut 6 3 3
```

where `3sym2adjnoteqcirc.aut` defines the automaton, consisting of the following lines:

```
3
1 22 33
2 11 33
3 11 22
```

Problem 32. Using the formula for the number of circular words in the previous problem, calculate the number of necklaces of length n composed of three colors where no two adjacent colors are the same. Find all necklaces of length $n = 6$.

Solution. The solution method is basically the same as for the Fibonacci and tribonacci necklaces that we calculated previously. We use the formula for $a(n)$ in the previous problem together with Burnside's Theorem to get the following formula for the number of necklaces.

$$b(n) = \frac{1}{n} \sum_{d|n} \phi(d) a(n/d) \tag{125}$$

The first ten values are

n	1	2	3	4	5	6	7	8	9	10
$b(n)$	0	3	2	6	6	14	18	36	58	108

This is sequence A106365 in the OEIS.

All the necklaces for $n = 6$ are shown below as a six sided polygon, white=0, gray=1 and black=2 (there are 14).

The example above is generated by combining the results of the following commands:

```
kneck 3 6 | autogrep 3sym2adjnoteqcirc_2.aut 1 1
kneck 3 6 | autogrep 3sym2adjnoteqcirc_2.aut 2 2
```

where `3sym2adjnoteqcirc_2.aut` defines the automaton, consisting of the following lines:

```
3
0 11 22
1 00 22
2 00 11
```

Problem 33. Using equation 16 for the number of necklaces of length n with m colors, find an expression for the number of necklaces of length 12 and 13 with m colors.

Solution. The divisors of twelve are $1, 2, 3, 4, 6, 12$ for which the totient values are $\phi(1) = \phi(2) = 1$, $\phi(3) = \phi(4) = \phi(6) = 2$, $\phi(12) = 4$. So from equation 16 we get

$$p_{12}(m) = \frac{1}{12}(m^{12} + m^6 + 2m^4 + 2m^3 + 2m^2 + 4m)$$

This increases very quickly with m

$$p_{12}(2) = 352,\ p_{12}(3) = 44368,\ p_{12}(4) = 1398500$$

The divisors of thirteen are $1, 13$ for which the totient values are $\phi(1) = 1$, $\phi(13) = 12$. So from equation 16 we get

$$p_{13}(m) = \frac{1}{13}(m^{13} + 12m)$$

$$p_{13}(2) = 632,\ p_{13}(3) = 122643,\ p_{13}(4) = 5162224$$

Problem 34. Derive equation 18 for the number of necklaces when n is a prime number.

Solution. When n is a prime, its only divisors are 1 and n, and these have totients $\phi(1) = 1$, $\phi(n) = n - 1$. So from equation 16 we get

$$p_n(m) = \frac{1}{n}(m^n + (n-1)m)$$

Problem 35. Fermat's little theorem says that if n is a prime and m is any integer then

$$m^n \equiv m \pmod{n}$$

In other words, m^n and m are congruent modulo n, meaning they both have the same remainder when divided by n. Use the necklace counting formulas to show that Fermat's little theorem must be true.

Solution. In the previous problem we showed that the number of necklaces with m colors and length n = prime is

$$\frac{1}{n}(m^n + (n-1)m)$$

which we can write as

$$\frac{m^n - m}{n} + m$$

Since the number of necklaces must be an integer, $m^n - m$ must be divisible by n which is another way of saying m^n and m must be congruent modulo n

$$m^n \equiv m \pmod{n}$$

There is another instructive way to look at this problem. Using m kinds of letters, it is possible to construct m^n different words of length n. Of these words, there will be m words where all letters are of a single kind. The number of words that contain at least 2 kinds of letters is then $m^n - m$. If n, the length of the words, is a prime, then no word can be expressed as a concatenation of copies of a smaller word, since the length of the smaller word would have to divide n, but n is prime. This means that no word has a periodic pattern and there must therefore be n other words related to it by a circular shift of the letters. The total number of these non-periodic words, $m^n - m$, must therefore be divisible by n which again gives us Fermat's little theorem. In terms of necklaces, we count the n circular shifted words as the same, so the total number of necklaces is

$$\frac{m^n - m}{n} + m$$

Problem 36. Find the number of bracelets of length n composed of 2 colors.

Solution. The solution is given by equation 27, with $m = 2$, in the Counting Bracelets section. The integer sequence is A000029 in the OEIS. The following table lists the first ten values.

n	1	2	3	4	5	6	7	8	9	10
$a(n)$	2	3	4	6	8	13	18	30	46	78

Problem 37. How many bracelets can you make with 5 black and 5 white beads?

Solution. The divisors of 10 are $[1, 2, 5, 10]$, and $\phi(1) = 1$, $\phi(2) = 1$, $\phi(5) = 4$, and $\phi(10) = 4$. Plugging these values into equation 26 of the Counting Bracelets section, we get the following cycle index polynomial for these bracelets

$$p(z) = \frac{1}{20}\left(z_1^{10} + z_2^5 + 4z_5^2 + 4z_{10}\right) + \frac{1}{4}\left(z_1^2 z_2^4 + z_2^5\right) \qquad (126)$$

Now make the substitution $z_k = b^k + 1$ in $p(z)$ and find the coefficient of b^5. The answer is 16.

All 16 bracelets are shown below, as a ten sided polygon.

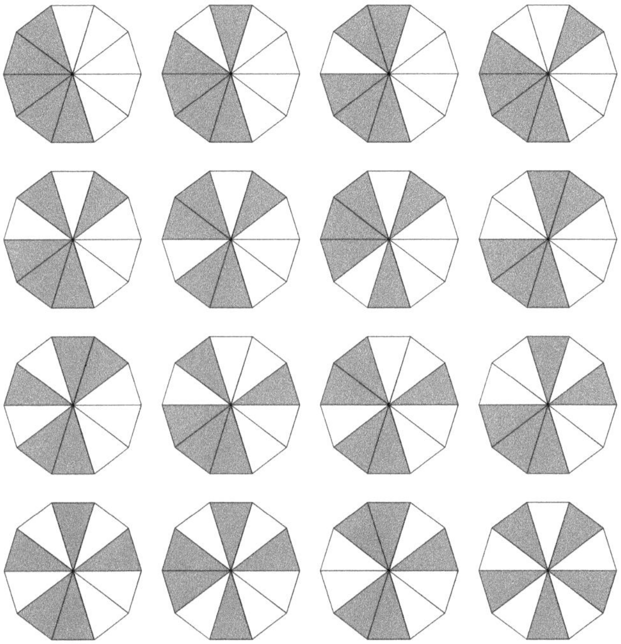

The example above is generated by the following command:

```
kbrace 2 10 | filterones.awk numones=5
```

Problem 38. Find the number of bracelets of length 4 composed of m colors where $m = 2, 3, \ldots 12$.

Solution. We can use equation 28, for the case of $p_4(m)$, in the Counting Bracelets section to solve this problem.

The table below lists the values. The table can be found as integer sequence A002817 in the OEIS.

m	2	3	4	5	6	7	8
$p_4(m)$	6	21	55	120	231	406	666

m	9	10	11	12
$p_4(m)$	1035	1540	2211	3081

All the length 4 bracelets for $m = 3$ are shown below as squares, white=0, gray=1 and black=2 (there are 21).

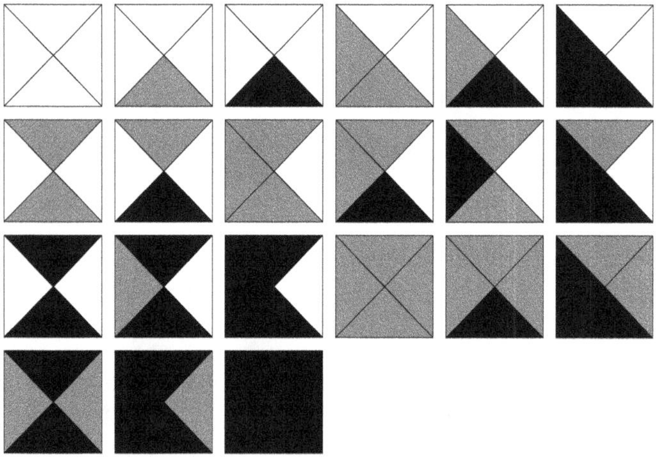

The example above is generated by the following command:

```
kbrace 3 4
```

Problem 39. Find the cycle index polynomial for a 2×2 square grid with respect to the group of $0°$, $90°$, $180°$, $270°$ rotations. Treat the rotations as permutations of the four squares of the grid as shown in figure 1.

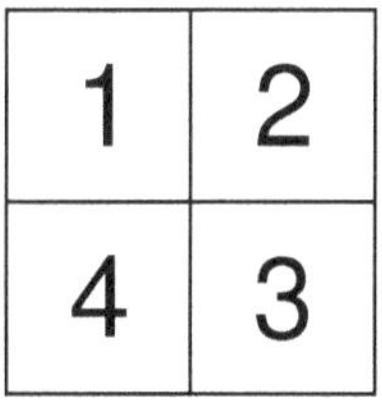

Figure 1: A 2×2 square grid.

Solution. The 2×2 square grid is equivalent to a necklace with 4 beads. We could just use the formula for the cycle index of a necklace with $n = 4$, but let's look at the permutations. In cycle notation they are

$0°$	$(1)(2)(3)(4)$	z_1^4
$90°$	$(1\ 2\ 3\ 4)$	z_4
$180°$	$(1\ 3)(2\ 4)$	z_2^2
$270°$	$(1\ 4\ 3\ 2)$	z_4

So the cycle index polynomial is

$$p(z) = \frac{1}{4}(z_1^4 + z_2^2 + 2z_4)$$

Doing the same thing with the formula, the divisors of 4 are $\{1, 2, 4\}$ with totients $\phi(1) = \phi(2) = 1$, $\phi(4) = 2$.

Plugging these values into the formula produces the same answer.

Problem 40. Find the cycle index polynomial for the 3×3 square grid shown in figure 2 with respect to the group of 0°, 90°, 180°, 270° rotations.

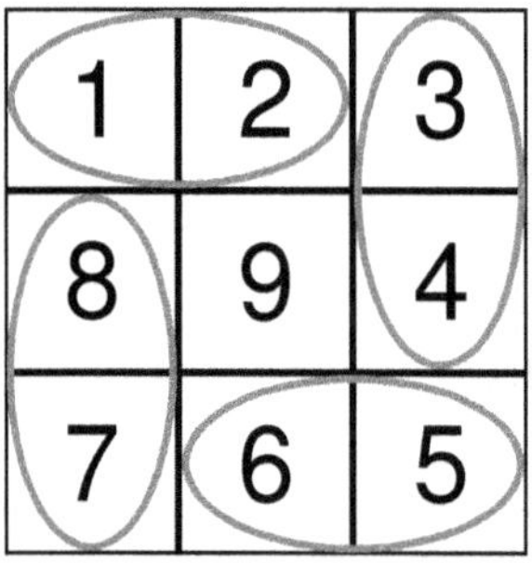

Figure 2: A 3×3 square grid as a 4 bead necklace with an extra bead in the center.

Solution. Figure 2 has been drawn in a way that suggests this is also a 4 bead necklace with an extra bead in the center. We start by looking at the permutations.

0°	$(1)(2)\ldots(9)$	z_1^9
90°	$(1\,3\,5\,7)(2\,4\,6\,8)(9)$	$z_4^2 z_1$
180°	$(1\,5)(2\,6)(3\,7)(4\,8)(9)$	$z_2^4 z_1$
270°	$(1\,7\,5\,3)(2\,8\,6\,4)(9)$	$z_4^2 z_1$

So the cycle index polynomial is

$$p(z) = \frac{1}{4} z_1 (z_1^8 + z_2^4 + 2z_4^2)$$

If the squares are grouped as $1-2$, $3-4$, $5-6$, $7-8$, then this becomes a 4 bead necklace problem with an extra bead in the center that is invariant with respect to the rotations. Grouping the squares in pairs means there are now two cycles for each cycle of the 4 bead necklace. Replace z_i with z_i^2 in the 4 bead cycle index, multiply by z_1 for the square in the center and you get the polynomial we found above.

Problem 41. Find the cycle index polynomial for the 4×4 square grid shown in figure 3 with respect to the group of $0°$, $90°$, $180°$, $270°$ rotations.

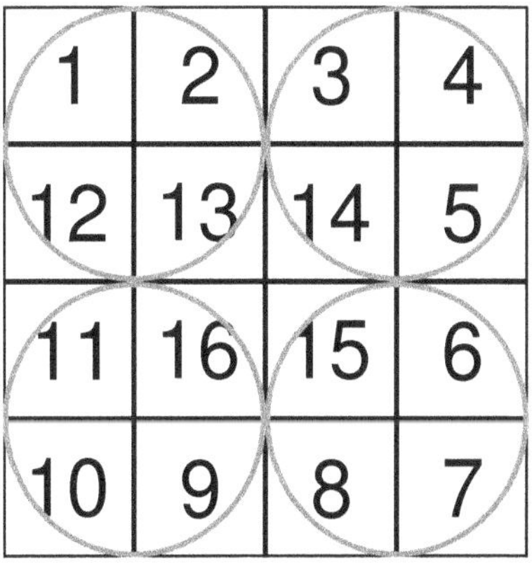

Figure 3: A 4×4 square grid as a 4 bead necklace with each bead composed of 4 sub-beads.

Solution. As shown in figure 3, the squares can be placed in four groups as $1-2-13-12$, $3-4-5-14$,

$6-7-8-15$, and $9-10-11-16$. This then becomes a 4 bead necklace problem where each bead is composed of 4 sub-beads. To get the cycle index, replace z_i with z_i^4 in the 4 bead cycle index.

$$p(z) = \frac{1}{4}(z_1^{16} + z_2^8 + 2z_4^4)$$

Problem 42. Find the cycle index polynomial for the general case of an $n \times n$ square grid when n is even. It should be found with respect to the group of 0°, 90°, 180°, 270° rotations.

Solution. When n is even, the squares can be grouped into four $(n/2) \times (n/2)$ square grids. We then have a four bead necklace problem with each bead composed of $n^2/4$ sub-beads. To get the cycle index, replace z_i with $z_i^{n^2/4}$ in the 4 bead cycle index.

$$p_n(z) = \frac{1}{4}(z_1^{n^2} + z_2^{n^2/2} + 2z_4^{n^2/4}) \quad n = \text{even} \tag{127}$$

Problem 43. Find the cycle index polynomial for the general case of an $n \times n$ square grid when n is odd. It should be found with respect to the group of 0°, 90°, 180°, 270° rotations.

Solution. When n is odd, the squares can be grouped into four $(n+1)/2 \times (n-1)/2$ rectangular grids, surrounding one square grid in the center, that remains invariant with respect to the rotations. We then have a four bead necklace problem with each bead composed of $(n^2-1)/4$ sub-beads. Replace z_i with $z_i^{(n^2-1)/4}$ in the 4 bead cycle index and multiply by z_1 for the invariant square in the center.

$$p_n(z) = \frac{1}{4} z_1 (z_1^{n^2-1} + z_2^{(n^2-1)/2} + 2z_4^{(n^2-1)/4}) \quad n = \text{odd} \tag{128}$$

Problem 44. How many ways can you color one of the squares of an $n \times n$ grid black and the others white under rotational symmetry?

Solution. The cycle index polynomials for even and odd n are equations 127 and 128, respectively. To find the number of colorings, we could replace z_i with $b^i + w^i$ and find the coefficient of the bw^{n^2-1} term in the resulting polynomial. But since the value of w is irrelevant, we can use $w = 1$ and replace z_i with $b^i + 1$. In the case where n is even, we have the polynomial

$$p_n(b) = \frac{1}{4}\left[(b+1)^{n^2} + (b^2+1)^{n^2/2} + 2(b^4+1)^{n^2/4}\right]$$

The coefficient of b in $(b+1)^{n^2}$ is n^2. The other binomials will not produce a b term. So the answer is $n^2/4$ for even n. When n is odd, we have the polynomial

$$p_n(b) = \frac{1}{4}(b+1)$$
$$\left[(b+1)^{n^2-1} + (b^2+1)^{(n^2-1)/2} + 2(b^4+1)^{(n^2-1)/4}\right]$$

$(b+1)^{n^2}$ will produce a b term with coefficient n^2, $(b+1)(b^2+1)^{(n^2-1)/2}$ will produce a b term with coefficient 1, and $2(b+1)(b^4+1)^{(n^2-1)/4}$ will produce a b term with coefficient 2. So the answer is $(n^2+3)/4$ for odd n. Summarizing, we have

$$f(n) = \begin{cases} n^2/4, & n = \text{even} \\ (n^2+3)/4, & n = \text{odd} \end{cases}$$

For various values of n we have

n	1	2	3	4	5	6	7	8	9	10
$f(n)$	1	1	3	4	7	9	13	16	21	25

The integer sequence is A004652 in the OEIS.

The squares for the case of a 3×3 grid are shown in Figure 4 (there are 3).

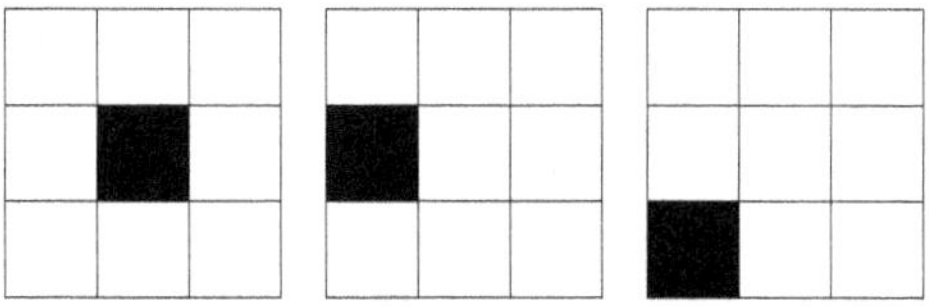

Figure 4: Ways to color one of the squares of a 3×3 grid black and the others white under rotational symmetry.

The example above is generated by the following commands:

```
kneck 4 4 | autogrep grid3.aut 0 0
kneck 4 4 | autogrep grid3.aut 0 1
```

where `grid3.aut` defines the automaton, consisting of the following lines:

```
5
0 00 11 21 32
1 01 12 22 33
2 02 13 23 34
3 03 14 24
4 04
```

The automaton is shown below.

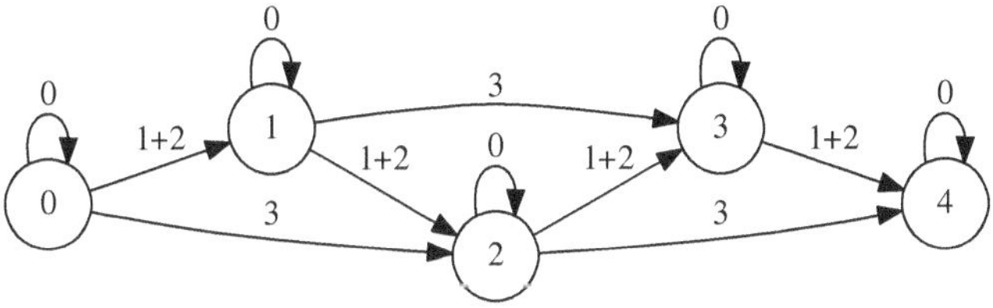

The four digit output of the commands is interpreted as a necklace with 4 double beads going clockwise from the top left of the 3×3 grid, around the perimeter. This takes care of the 8 outer squares of the grid, with the color of the center square determined by how many 1's (black squares) have already occurred.

Problem 45. How many ways can you color two of the squares of an $n \times n$ grid black and the others white under rotational symmetry?

Solution. The only difference between this and problem 9 is that now we want the coefficient of b^2. For n even, the coefficient of b^2 in $(b+1)^{n^2}$ is $\binom{n^2}{2} = \frac{1}{2}n^2(n^2-1)$. The coefficient of b^2 in $(b^2+1)^{n^2/2}$ is $\frac{1}{2}n^2$. So for even n, the number of ways is $\frac{1}{4}\left[\frac{1}{2}n^2(n^2-1)+\frac{1}{2}n^2\right] = \frac{1}{8}n^4$. For n odd, $(b+1)^{n^2}$ will produce a b^2 term with coefficient $\binom{n^2}{2}$. $(b+1)(b^2+1)^{(n^2-1)/2}$ will produce a b^2 term with coefficient $\frac{1}{2}(n^2-1)$. $2(b+1)(b^4+1)^{(n^2-1)/4}$ will not produce a b^2 term. So for odd n, the number of ways is $\frac{1}{4}\left[\frac{1}{2}n^2(n^2-1)+\frac{1}{2}(n^2-1)\right] = \frac{1}{8}(n^2-1)(n^2+1)$. Summarizing, we have

$$f(n) = \begin{cases} \frac{1}{8}n^4, & n = \text{even} \\ \frac{1}{8}(n^2-1)(n^2+1), & n = \text{odd} \end{cases}$$

For various values of n we have

n	1	2	3	4	5	6	7	8	9	10
$f(n)$	0	2	10	32	78	162	300	512	820	1250

The integer sequence is A212714 in the OEIS.

The squares for the case of a 3×3 grid are shown in Figure 5 (there are 10).

The example above is generated by the following commands:

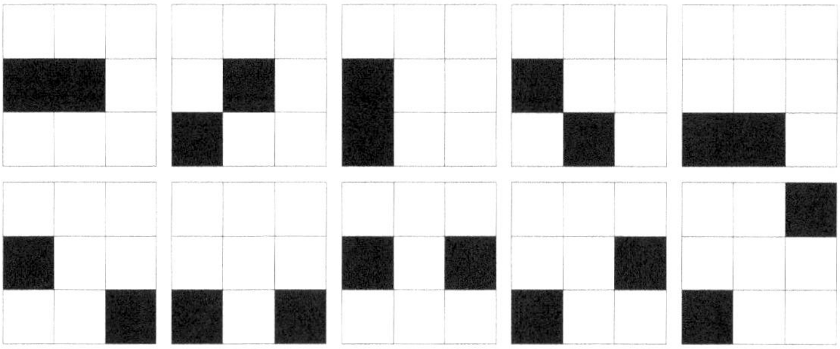

Figure 5: Ways to color two of the squares of a 3×3 grid black and the others white under rotational symmetry.

```
kneck 4 4 | autogrep grid3.aut 0 1
kneck 4 4 | autogrep grid3.aut 0 2
```

where `grid3.aut` defines the automaton listed previously.

See Problem 44 for how the output of the above two commands should be interpreted.

Problem 46. Same as the previous problem, but now we want to color three of the squares black.

Solution. The number of ways to color three of the squares black will come from the coefficient of b^3. For n even, only $(b+1)^{n^2}$ will produce a b^3 term, and the coefficient is $\binom{n^2}{3}$. So for even n, the number of ways is $\frac{1}{4}\binom{n^2}{3}$. For n odd, we have $\binom{n^2}{3}$ from $(b+1)^{n^2}$ and

$\frac{1}{2}(n^2-1)$ from $(b+1)(b^2+1)^{(n^2-1)/2}$. So for odd n, the number of ways is $\frac{1}{4}\left[\binom{n^2}{3}+\frac{1}{2}(n^2-1)\right]$. Summarizing, we have

$$f(n) = \begin{cases} \frac{1}{4}\binom{n^2}{3}, & n = \text{even} \\ \frac{1}{4}\left[\binom{n^2}{3}+\frac{1}{2}(n^2-1)\right], & n = \text{odd} \end{cases}$$

For the first few values of n we have

n	1	2	3	4	5	6	7	8
$f(n)$	0	1	22	140	578	1785	4612	10416

The integer sequence is A275799 in the OEIS.

The squares for the case of a 3×3 grid are shown in Figure 6 (there are 22).

The example above is generated by the following commands:

```
kneck 4 4 | autogrep grid3.aut 0 2
kneck 4 4 | autogrep grid3.aut 0 3
```

where `grid3.aut` defines the automaton listed previously.

See Problem 44 for how the output of the above two commands should be interpreted.

Problem 47. Same as the previous problem, but four of the squares are black.

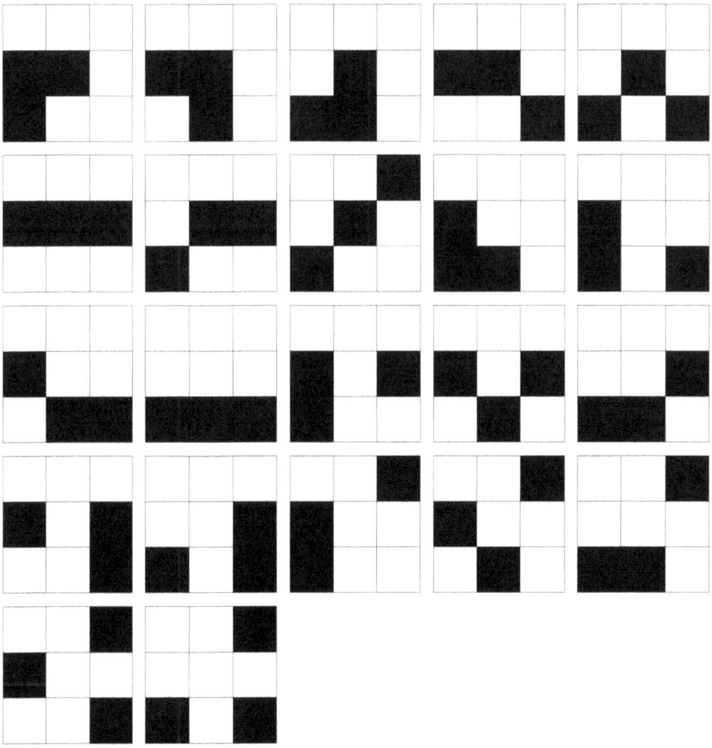

Figure 6: Ways to color three of the squares of a 3×3 grid black and the others white under rotational symmetry.

Solution. The solution is the coefficient of b^4. For n even, $(b+1)^{n^2}$ contributes $\binom{n^2}{4}$, $(b^2+1)^{n^2/2}$ contributes $\binom{n^2/2}{2}$, and $2(b^4+1)^{n^2/4}$ contributes $n^2/2$. The number of ways for even n is $\frac{1}{4}\left[\binom{n^2}{4}+\binom{n^2/2}{2}+\frac{1}{2}n^2\right]$. For n odd, $(b+1)^{n^2}$ contributes $\binom{n^2}{4}$, $(b+1)(b^2+1)^{(n^2-1)/2}$ contributes $\binom{(n^2-1)/2}{2}$, and $2(b+1)(b^4+1)^{(n^2-1)/4}$ contributes $\frac{1}{2}(n^2-1)$. The number of ways for odd n is $\frac{1}{4}\left[\binom{n^2}{4}+\binom{(n^2-1)/2}{2}+\frac{1}{2}(n^2-1)\right]$.

Summarizing, we have

$$f(n)=\frac{1}{4}\begin{cases}\binom{n^2}{4}+\binom{n^2/2}{2}+\frac{1}{2}n^2, & n=\text{even}\\ \binom{n^2}{4}+\binom{(n^2-1)/2}{2}+\frac{1}{2}(n^2-1), & n=\text{odd}\end{cases}$$

For the first eight values of n we have

n	1	2	3	4	5	6	7	8
$f(n)$	0	1	34	464	3182	14769	53044	158976

The integer sequence is A277226 in the OEIS.

The squares for the case of a 3×3 grid are shown in Figure 7 (there are 34).

The example above is generated by the following commands:

```
kneck 4 4 | autogrep grid3.aut 0 3
kneck 4 4 | autogrep grid3.aut 0 4
```

Figure 7: Ways to color four of the squares of a 3×3 grid black and the others white under rotational symmetry.

where `grid3.aut` defines the automaton listed previously.

See Problem 44 for how the output of the above two commands should be interpreted.

Problem 48. Extend the previous four problems to the case of k black squares.

Solution. As in the previous problems, the solution is the coefficient of b^k. We need to look at four situations. For even n and even k, $(b+1)^{n^2}$ contributes $\binom{n^2}{k}$, $(b^2+1)^{n^2/2}$ contributes $\binom{n^2/2}{k/2}$, and $2(b^4+1)^{n^2/4}$ contributes $2\binom{n^2/4}{k/4}$ when k is a multiple of 4, and 0 otherwise. For even n and even k, the answer is

$$f_k(n) = \frac{1}{4}\begin{cases} \binom{n^2}{k} + \binom{n^2/2}{k/2} + 2\binom{n^2/4}{k/4}, & k = 0, 4, 8, \ldots \\ \binom{n^2}{k} + \binom{n^2/2}{k/2}, & k = 2, 6, 10, \ldots \end{cases} \tag{129}$$

For even n and odd k, the only contribution is $\binom{n^2}{k}$ from $(b+1)^{n^2}$ so the answer is

$$f_k(n) = \frac{1}{4}\binom{n^2}{k} \tag{130}$$

For odd n and even k, $(b+1)^{n^2}$ contributes $\binom{n^2}{k}$, $(b+1)(b^2+1)^{(n^2-1)/2}$ contributes $\binom{(n^2-1)/2}{k/2}$, and $2(b+1)(b^4+$

$1)^{(n^2-1)/4}$ contributes $2\binom{(n^2-1)/4}{k/4}$ when k is a multiple of 4, and 0 otherwise. For odd n and even k, the answer is

$$f_k(n) = \frac{1}{4}\begin{cases} \binom{n^2}{k} + \binom{(n^2-1)/2}{k/2} + 2\binom{(n^2-1)/4}{k/4}, & k = 0, 4, 8, \ldots \\ \binom{n^2}{k} + \binom{(n^2-1)/2}{k/2}, & k = 2, 6, 10, \ldots \end{cases} \tag{131}$$

For odd n and odd k, $(b+1)^{n^2}$ contributes $\binom{n^2}{k}$, $(b+1)(b^2+1)^{(n^2-1)/2}$ contributes $\binom{(n^2-1)/2}{(k-1)/2}$, and $2(b+1)(b^4+1)^{(n^2-1)/4}$ contributes $2\binom{(n^2-1)/4}{(k-1)/4}$ when k is a multiple of 4, and 0 otherwise. For odd n and odd k, the answer is

$$f_k(n) = \frac{1}{4}\begin{cases} \binom{n^2}{k} + \binom{(n^2-1)/2}{(k-1)/2} + 2\binom{(n^2-1)/4}{(k-1)/4}, & k = 0, 4, 8, \ldots \\ \binom{n^2}{k} + \binom{(n^2-1)/2}{(k-1)/2}, & k = 2, 6, 10, \ldots \end{cases} \tag{132}$$

Problem 49. A strongly binary tree (SBT) is a rooted tree where each vertex, including the root, has zero or two children. Figure 8 shows an example of an SBT with 5 vertices. The root is labeled 0 and it has two children labeled 1 and 2. Vertex 1 has no children, and vertex 2 has two children labeled 3 and 4. Vertices 1, 3, and 4 are called the leaves of the tree. The tree is symmetric in the sense that if it is reflected about a vertical axis through the root (vertex 0) then it is still considered to be the same tree. Figure 9 shows all the

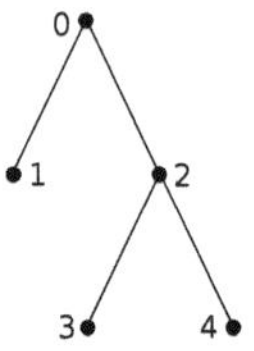

Figure 8: SBT with 5 vertices.

SBT's with $n = 1, 3, 5, 7$ vertices. The problem is to find the number of unique SBT's with n vertices.

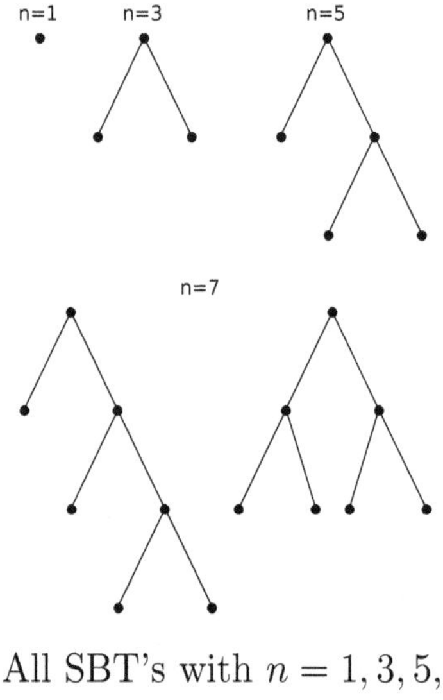

Figure 9: All SBT's with $n = 1, 3, 5, 7$ vertices.

Solution. The reflection symmetry suggests that we can solve this problem using the cycle index polynomial for the permutation group S_2. The group contains only two permutations, (1)(2) and (1 2) so the cycle index is

$$p(z) = \frac{1}{2}(z_1^2 + z_2)$$

To get the pattern inventory, we make the substitutions

$$z_1 \to s(t) \quad z_2 \to s(t^2)$$

where $s(t)$ is the generating function for the number of SBT's, that is $s(t)$ is a formal power series of the following form

$$s(t) = \sum_{n=1}^{\infty} s_n t^n$$

where the coefficient of t^n, s_n, is the number of SBT's with n vertices. This is just like the substitutions that were used in previous problems such as counting necklaces. What we have here, essentially, is a two bead necklace problem where each bead can be composed of n sub-beads (vertices), all of the same kind, that can be arranged in s_n ways. Making the substitution, we get

$$p(t) = \frac{1}{2}(s^2(t) + s(t^2))$$

The coefficient of t^n in this expression is the number of ways to combine two SBT's under reflection symmetry (switching left and right makes no difference) so that the total number of vertices in the combined SBT's is

n. The key to solving the problem is to note that if the two SBT's are combined by introducing a new root node that attaches to their root nodes, then the result is a new SBT. Figure 10 shows the idea. The dashed

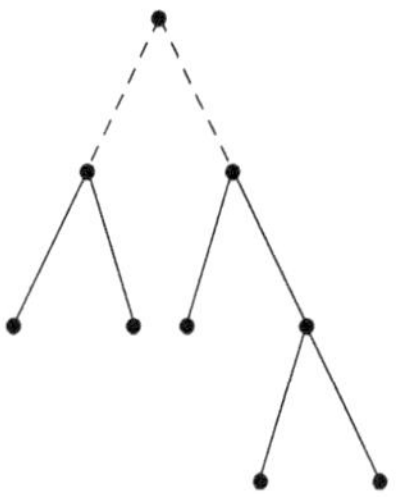

Figure 10: Combining two SBT's.

line shows the new root node attached to an SBT with 3 vertices and one with 5 vertices. The new combined SBT has $3 + 5 + 1 = 9$ vertices. The new root node is taken into account by multiplying $p(t)$ by t (recall that each t represents a node or vertex). The coefficient of t^n in $tp(t)$ is then equal to the number of SBT's with n vertices. If we add t for the tree with a single vertex, then we just get $s(t)$

$$tp(t) + t = s(t)$$

This gives us the following functional equation for $s(t)$

$$\frac{1}{2}t(s^2(t) + s(t^2)) + t = s(t)$$

This equation can be used to solve for the coefficients, s_n, of the power series for $s(t)$. From figure 9, we know that $s_1 = s_3 = s_5 = 1$ and $s_7 = 2$. It is also clear that, since we have one root vertex and every vertex can have only 0 or 2 children, the number of vertices must be odd. So $s(t)$ will have the following form.

$$s(t) = t + t^3 + t^5 + 2t^7 + s_9 t^9 + \ldots$$

The coefficient of t^{2n+1} in $ts^2(t)$ is given by

$$c_n = \sum_{k=0}^{n-1} s_{2k+1} s_{2n-(2k+1)} \qquad n = 1, 2, 3, \ldots$$

Putting this together with $ts(t^2)$, we get the following equation for the coefficients, $n = 1, 2, 3, \ldots$

$$s_{2n+1} = \frac{1}{2} \begin{cases} c_n + s_n, & n = \text{odd} \\ c_n, & n = \text{even} \end{cases}$$

The equation can be simplified somewhat, but we leave it in this form. The calculation for s_9 and s_{11} is

$$\begin{aligned} s_9 &= \frac{1}{2} c_4 \\ &= \frac{1}{2}(s_1 s_7 + s_3 s_5 + s_5 s_3 + s_7 s_1) = 3 \end{aligned}$$

$$\begin{aligned} s_{11} &= \frac{1}{2}(c_5 + s_5) \\ &= \frac{1}{2}(s_1 s_9 + s_3 s_7 + s_5 s_5 + s_7 s_3 + s_9 s_1 + s_5) = 6 \end{aligned}$$

Some values of s_n are

n	0	1	2	3	4	5	6	7	8	9	10
$2n+1$	1	3	5	7	9	11	13	15	17	19	21
s_{2n+1}	1	1	1	2	3	6	11	23	46	98	207

This is sequence A001190 in the OEIS. They are called the Wedderburn-Etherington numbers.

Problem 50. A general binary tree (BT) is a rooted tree where each vertex, including the root, can have 0, 1, or 2 children. At each vertex, we say there is a right and left branch, corresponding to the 2 possible children. Either or both of the branches may be empty, depending on the number of children. The tree has no symmetry in the sense that switching the right and left branch of any vertex produces a different tree. Figure 11 shows all possible trees with three vertices. All 5 of

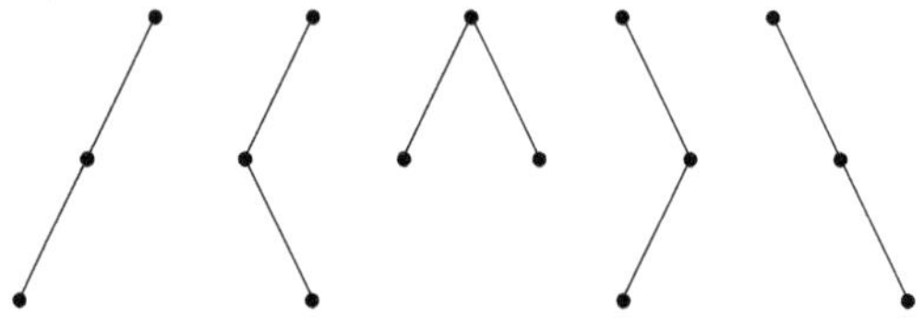

Figure 11: All possible general binary trees with 3 vertices.

these trees are considered unique. The problem is to find the number of binary trees with n vertices.

Solution. The key to solving this problem is to note that every binary tree is composed of two binary trees corresponding to the right and left branch of the root. We treat these two trees as the beads of a two bead necklace that has only the identity as a symmetry, i.e. (1)(2) is the only permutation. The cycle index is then

$$p(z) = z_1^2 \tag{133}$$

The pattern inventory is found by making the substitution $z_1 \to b(t)$ where $b(t)$ is the generating function for the number of binary trees. It is a power series of the following form

$$b(t) = \sum_{n=0}^{\infty} b_n t^n \tag{134}$$

where b_n is the number of binary trees with n vertices. With this substitution, the pattern inventory becomes

$$p(t) = b^2(t) \tag{135}$$

The coefficient of t^n in $p(t)$ is the number of ways to select two binary trees, for the right and left branch of the root, such that the combined number of vertices in the two branches is n. To add a root node to this, we multiply by t so that $tp(t)$ is then the generating function for the number of binary trees with at least one vertex (the root). If we add 1 for the null tree, then $tp(t) + 1 = b(t)$, the generating function for the number of binary trees. So we get the following functional equation for $b(t)$.

$$tb^2(t) + 1 = b(t) \tag{136}$$

These are two ways to get the values of b_n from this equation. We can get a recurrence equation for b_n directly from the equation or we can solve for $b(t)$ and get the b_n values from that. The recurrence equation is fairly easy to get in this case. The coefficient of t^n in $tb^2(t)$ is given by a convolution that should equal the coefficient b_n on the right side of equation 136, so we have

$$b_n = \sum_{k=0}^{n-1} b_k b_{n-1-k}$$

With $b_0 = 1$, we get

$$\begin{aligned} b_1 &= b_0^2 = 1 \\ b_2 &= 2b_0b_1 = 2 \\ b_3 &= 2b_0b_2 + b_1^2 = 5 \\ b_4 &= 2(b_0b_3 + b_1b_2) = 14 \\ b_5 &= 2(b_0b_4 + b_1b_3) + b_2^2 = 42 \end{aligned}$$

We can also get these values by first solving for $b(t)$ in equation 136.

$$b(t) = \frac{1 \pm \sqrt{1-4t}}{2t}$$

Only the minus sign solution will produce positive values for b_n, so we use

$$b(t) = \frac{1 - \sqrt{1 - 4t}}{2t}$$

The binomial theorem then gives us the following equation for b_n.

$$b_n = \frac{1}{n+1}\binom{2n}{n}$$

The first few values of b_n are

n	0	1	2	3	4	5	6	7	8	9
b_n	1	1	2	5	14	42	132	429	1430	4862

This sequence is A000108 in the OEIS. It is a well known sequence called the Catalan numbers.

Problem 51. A general trinary tree is a rooted tree where each vertex, including the root, can have 0, 1, 2, or 3 children. At each vertex, there is a right, left, and center branch corresponding to the 3 possible children. Any or all of the branches may be empty. The tree has no symmetry, in the sense that any permutation of the branches of any vertex results in a different tree. Figure 12 shows all 12 unique trinary trees with 3 vertices. The problem is to find the number of trinary trees with n vertices.

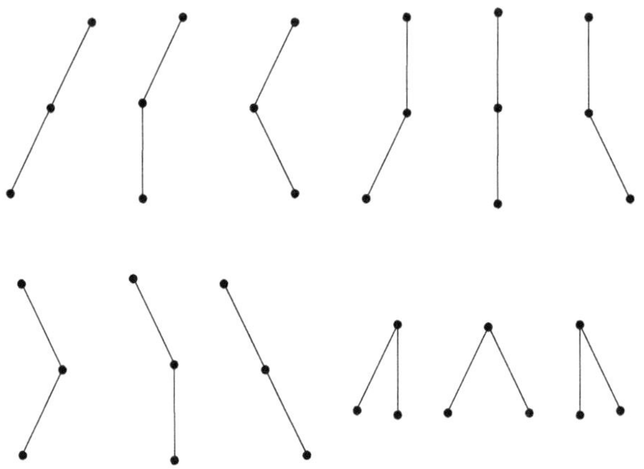

Figure 12: All 12 general trinary trees with 3 vertices.

Solution. We will use the same approach as the previous problem where we looked at binary trees. Every trinary tree is composed of three trinary trees corresponding to the right, left, and center branches of the root. These are like the three beads of a necklace with only the identity (1)(2)(3) as a symmetry. So the cycle index is

$$p(z) = z_1^3$$

We get the pattern inventory using the substitution $z_1 \to c(t)$ where $c(t)$ is the generating function for the number of trinary trees.

$$c(t) = \sum_{n=0}^{\infty} c_n t^n$$

and c_n is the number of trinary trees with n vertices. The pattern inventory is then

$$p(t) = c^3(t)$$

As in the binary tree problem, if we multiply this by t for the root node and add 1 for the null tree, we get $tp(t) + 1 = c(t)$. So we have the following functional equation for $c(t)$.

$$tc^3(t) + 1 = c(t)$$

The coefficient of t^n in $tc^3(t)$ is given by a double convolution and it is equal to c_n on the right side of the equation. Working this out for the first few terms, with $c_0 = 1$, we have

$$\begin{aligned}
c_1 &= c_0^3 = 1 \\
c_2 &= 3c_0^2c_1 = 3 \\
c_3 &= 3(c_0c_1^2 + c_0^2c_2) = 12 \\
c_4 &= c_1^3 + 6c_0c_1c_2 + 3c_0^2c_3 = 55 \\
c_5 &= 3(c_1^2c_2 + c_0c_2^2) + 2c_0c_1c_3 + c_0^2c_4) = 273
\end{aligned}$$

The first few values are

n	0	1	2	3	4	5	6	7	8
c_n	1	1	3	12	55	273	1428	7752	43263

This sequence is A001764 in the OEIS. You can verify that the formula for c_n is

$$c_n = \frac{1}{2n+1}\binom{3n}{n}$$

Problem 52. A tetrahedron is one of the five Platonic solids. These are polyhedra where all the faces are the same convex regular polygon and the same number of faces meet at each vertex. A tetrahedron has four faces that are equilateral triangles, and three of them meet at each vertex. It is a pyramid whose base and three sides are equilateral triangles. Figure 13 shows an apex view of the tetrahedron. There are four vertices

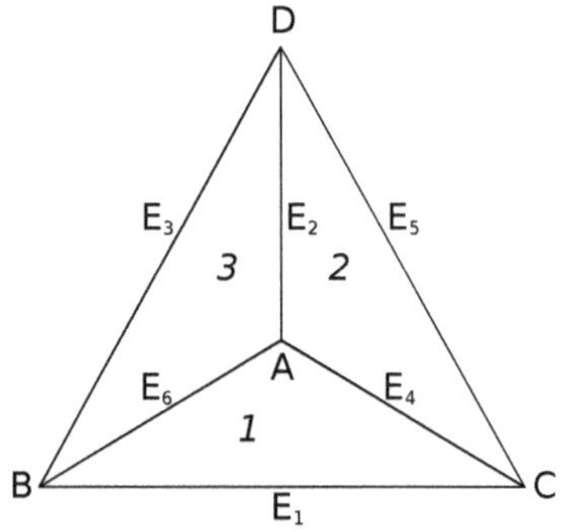

Figure 13: Tetrahedron apex view.

labeled A, B, C, D, and six edges labeled E_1, E_2, E_3, E_4, E_5, E_6. The problem is to find the number of unique ways a tetrahedron can be colored with m colors. The colorings must be unique with respect to all possible rotations, i.e. if one coloring can be gotten from another by a rotation, then the two colorings are considered equivalent.

Solution. The first step is to find how the faces are permuted under all possible rotations. Figure 14 shows the tetrahedron with the sides folded down flat onto the plane. The four faces are labeled 1, 2, 3, 4. For each of

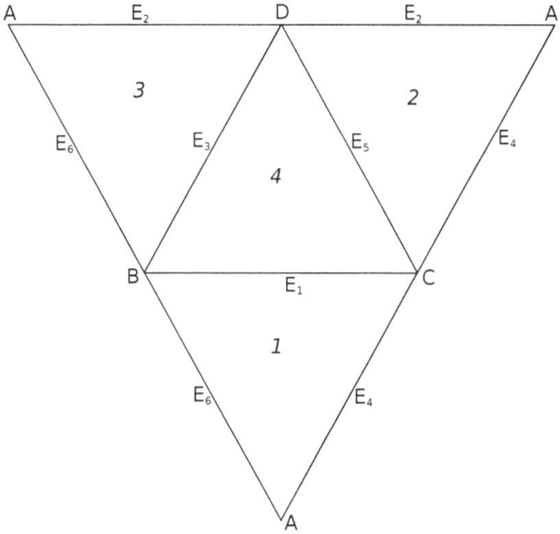

Figure 14: Tetrahedron folded down.

the vertices, there is a rotation axis between it and the center of the opposite face. There are 120° and 240° rotations about these axes. For the axis through vertex A, the rotations have the following cycle structure.

$$\begin{aligned} 120^\circ &: (1\ 2\ 3)(4) \\ 240^\circ &: (1\ 3\ 2)(4) \end{aligned}$$

Rotations about the other vertices will have similar cycle structure. All the vertex rotations together will contribute $8z_1z_3$ to the cycle index polynomial. There are three more rotation axes through the centers of edge pairs $E_1 - E_2$, $E_3 - E_4$, $E_5 - E_6$. There is a single 180° rotation about each of these axes. The rotation about the $E_1 - E_2$ axis has cycle structure (1 4)(2 3). The other edge axes have the same structure, so all these rotations contribute $3z_2^2$ to the cycle index polynomial. There are a total of 12 rotational symmetries including the identity, so the cycle index polynomial is

$$p(z) = \frac{1}{12}(z_1^4 + 3z_2^2 + 8z_1z_3) \tag{137}$$

To get the number of colorings, substitute $z_i \rightarrow m$.

$$p(m) = \frac{1}{12}(m^4 + 11m^2)$$

For the first 10 values of m, we have

m	1	2	3	4	5	6	7	8	9	10
$p(m)$	1	5	15	36	75	141	245	400	621	925

The integer sequence is A006008 in the OEIS.

Problem 53. How many unique ways can you color the four vertices of a tetrahedron, where, as in the previous problem, the uniqueness is with respect to rotations.

Solution. The answer is the same as the number of ways to color the faces that was worked out in the previous problem. To see this, note that for a given tetrahedron, the center of each of its four faces is the vertex of a smaller tetrahedron. For a tetrahedron, the faces and the vertices have the same set of rotational symmetries. In this respect, the tetrahedron is said to be self-dual. Two polyhedra are said to be duals of each other if the faces of one correspond to the vertices of the other, and vice versa.

Problem 54. How many unique ways, under rotation, can you color the four faces of a tetrahedron with four colors, such that each of the colors is used?

Solution. Let the four colors be symbolized by R, G, B, and Y. Then we make the substitution $z_i \to R^i + G^i + B^i + Y^i$ into the cycle index polynomial in equation 137 and find the coefficient of $RGBY$. The only place where this term appears is in $z_1^4 \to (R+G+B+Y)^4$ with coefficient equal to 24. Dividing this by 12, we get the answer 2. There are 2 unique ways to color the faces with all four colors.

Problem 55. How many unique ways, under rotation, can you color two of the faces of a tetrahedron black and the other two white?

Solution. The answer is found by making the substitution $z_i \to B^i + 1$ into the cycle index polynomial in equation 137 and finding the coefficient of B^2. The cycle index polynomial becomes

$$p(B) = \frac{1}{12}((B+1)^4 + 3(B^2+1)^2 + 8(B+1)(B^3+1))$$

which simplifies to

$$p(B) = B^4 + B^3 + B^2 + B + 1$$

So the answer is 1.

Problem 56. The most familiar of the Platonic solids is probably the cube. It has six square faces, eight vertices and twelve edges. Figure 15 shows a view of the cube looking slightly above one of the diagonals. The vertices are labeled $V_1 - V_8$, the edges $E_1 - E_{12}$, and the faces $1-6$. The problem is to find the number of unique ways the faces of a cube can be colored with m colors. The colorings must be unique with respect to rotations.

Solution. To solve the problem, we need to find how the faces are permuted under all possible rotations. There are three rotation axes through the centers of opposite faces. The possible rotations are 90°, 180°, 270°. For the axis through the center of face 5 and 6, the rotations have the following cycle structure.

$$\begin{aligned} 90^\circ &: (1\ 2\ 3\ 4)(5)(6) \\ 180^\circ &: (1\ 3)(2\ 4)(5)(6) \\ 270^\circ &: (1\ 4\ 3\ 2)(5)(6) \end{aligned}$$

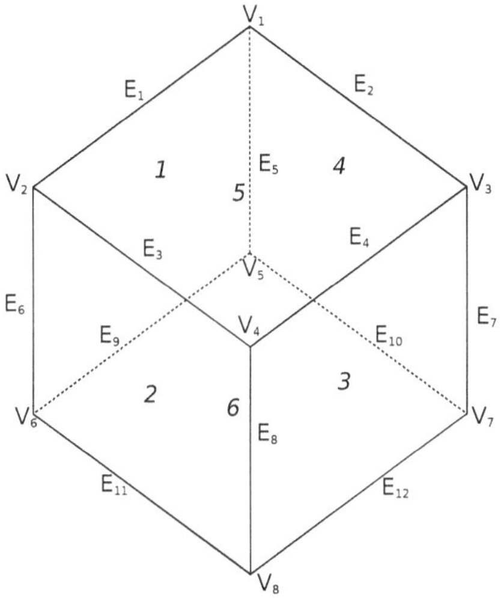

Figure 15: The cube looking along a diagonal.

The other face rotations will have a similar cycle structure so all together these rotations will contribute

$6z_1^2 z_4 + 3z_1^2 z_2^2$

to the cycle index polynomial. There are four rotation axes along the four diagonals with possible rotations of 120° and 240°. For rotations about the axis through vertices V_4 and V_5, we have the following cycle structure.

$$\begin{aligned} 120^\circ &: (5\ 2\ 3)(1\ 6\ 4) \\ 240^\circ &: (5\ 3\ 2)(1\ 4\ 6) \end{aligned}$$

The other diagonal rotations have the same cycle structure, so all together they contribute $8z_3^2$ to the cycle index polynomial.

There are six rotation axes through the centers of opposite edges with a rotation of 180°. The cycle structure for the rotation through the center of E_5 and E_8 is (1 4)(2 3)(5 6). Altogether these rotations contribute $6z_2^3$.

The cycle index polynomial is then

$$p(z) = \frac{1}{24}(z_1^6 + 6z_1^2 z_4 + 3z_1^2 z_2^2 + 8z_3^2 + 6z_2^3) \tag{138}$$

To get the number of colorings, substitute $z_i \to m$.

$$p(m) = \frac{1}{24}(m^6 + 3m^4 + 12m^3 + 8m^2)$$

For the first 10 values of m, we have

m	1	2	3	4	5	6	7
$p(m)$	1	10	57	240	800	2226	5390

m	8	9	10
$p(m)$	11712	23355	43450

The integer sequence is A047780 in the OEIS.

Problem 57. How many unique ways, under rotation, can you color the faces of a cubic die black and white such that when the die is rolled, there is an equal probability of getting white or black?

Solution. There are six faces, so to get an equal probability, three of them must be colored white, and the other three black. To find the number of ways to do this, make the substitution, $z_i \to B^i+1$ in equation 138 and find the coefficient of B^3. Making the substitution and simplifying, we get

$$p(B) = B^6 + B^5 + 2B^4 + 2B^3 + 2B^2 + B + 1$$

So the answer is 2. There are also 2 unique ways to have 4 black faces and 2 white faces, and by symmetry, there are 2 unique ways to have 2 black faces and 4 white faces.

Problem 58. How many unique ways, under rotation, can you label the six faces of a cube with the numbers 1 through 6 such that each number is used?

Solution. Let the six numbers be symbolized by $x_1, x_2, \ldots x_6$. Then we make the substitution $x_i \to x_1^i + x_2^i + \cdots + x_6^i$ into the cycle index polynomial in equation 138 and find the coefficient of $x_1 x_2 \cdots x_6$. The only place where this term appears is in $z_1^6 \to (x_1 + x_2 + \cdots + x_6)^6$ with coefficient equal to 720. Dividing this by 24, we get the answer 30. There are 30 unique ways to label the faces of a cube with all the numbers 1 through 6. This is the number of ways you can manufacture the dice that are used in games of chance. Gambling dice are almost always manufactured so that the numbers on opposite faces add up to 7. For this to be possible, the faces numbered 1, 2 and 3 must meet at a vertex. Looking down at the vertex, the faces can be numbered clockwise or counterclockwise. So there are two ways to number dice so that numbers on opposite faces add up to 7. The two ways are called left handed and right handed dice. In left handed dice, the 1, 2 and 3 faces are labeled clockwise around a vertex, and in right handed dice, they are labeled counterclockwise. Dice used in the United States are usually right handed (see the Wikipedia article on dice at https://en.wikipedia.org/wiki/Dice). The two ways that gambling dice are labeled are shown below.

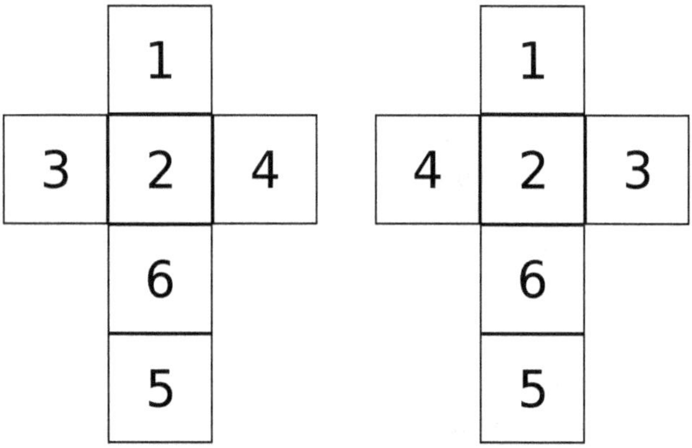

Figure 16: The two ways gambling dice are labeled.

Problem 59. The octahedron is a Platonic solid with eight faces, six vertices, and twelve edges. Each face is an equilateral triangle. Figures 17, 18, and 19 show an octahedron looking into a face, vertex, and edge, respectively. An octahedron can be constructed from two square base pyramids that have equilateral triangles as sides. The octahedron and the cube are duals of each other. The faces of the octahedron correspond to the vertices of the cube, and the faces of the cube correspond to the vertices of the octahedron. The problem is to find the number of unique ways the faces of an octahedron can be colored with m colors. The colorings must be unique with respect to rotations.

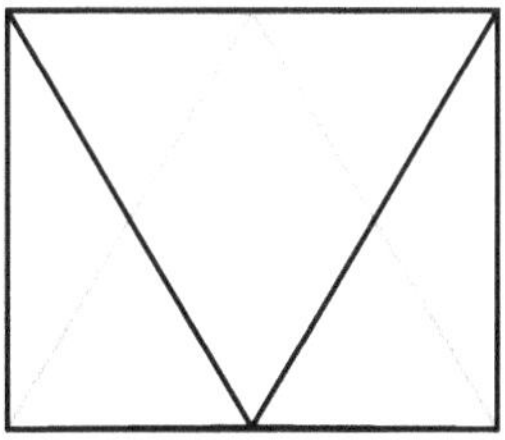

Figure 17: Octahedron face view.

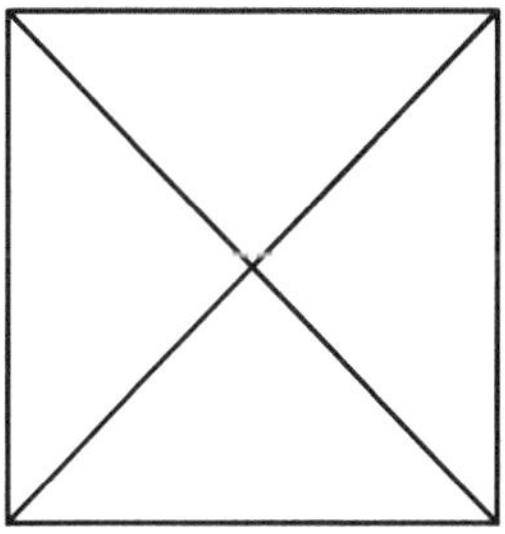

Figure 18: Octahedron vertex view.

Solution. As in the previous problems, we need to find how the faces are permuted under all possible ro-

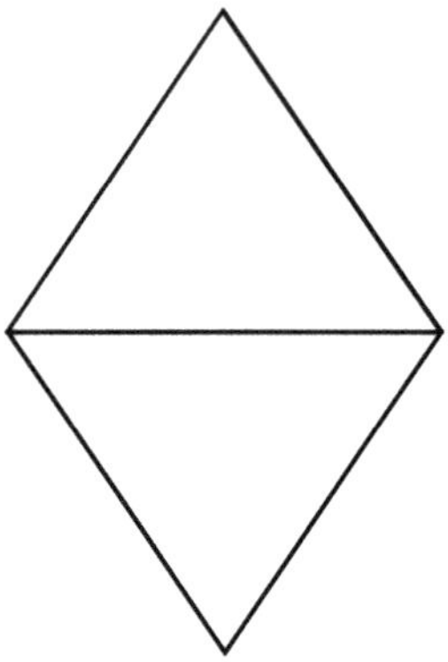

Figure 19: Octahedron edge view.

tations. There are four rotation axes through the centers of opposite faces, with possible rotations of 120°, and 240°. Each of these rotations fixes two of the faces, and the two sets of three faces adjacent to the two fixed faces are cyclically permuted. The cycle structure will then look like (1)(2)(3 4 5)(6 7 8) where faces 1 and 2 are fixed, and the two cycles of length three are the faces adjacent to them. All these face rotations contribute $8z_1^2z_3^2$ to the cycle index.

There are three rotation axes through opposite pairs of vertices with rotations of 90°, 180°, and 270°. These rotations permute the four faces that meet at each vertex. The cycle structure for each rotation is

$$\begin{aligned} 90^\circ &: (1\,2\,3\,4)(5\,6\,7\,8) \\ 180^\circ &: (1\,3)(2\,4)(5\,7)(6\,8) \\ 270^\circ &: (1\,4\,3\,2)(5\,8\,7\,6) \end{aligned}$$

All these vertex rotations together contribute $6z_4^2 + 3z_2^4$ to the cycle index.

There are six rotation axes through the centers of opposite pairs of edges with a possible rotation of 180°. The rotations permute the faces in pairs, and there are four such pairs for each rotation. All these edge rotations together contribute $6z_2^4$ to the cycle index.

Including the identity, we have a total of 24 rotational symmetries, and the cycle index polynomial is

$$p(z) = \frac{1}{24}(z_1^8 + 9z_2^4 + 8z_1^2 z_3^2 + 6z_4^2) \tag{139}$$

To get the number of colorings, substitute $z_i \to m$.

$$p(m) = \frac{1}{24}(m^8 + 17m^4 + 6m^2)$$

For the first 6 values of m, we have

m	1	2	3	4	5	6
$p(m)$	1	23	333	2916	16725	70911

The integer sequence is A000543 in the OEIS.

Problem 60. How many unique ways, under rotation, can you color the faces of an octahedron black and white such that there are an equal number of white and black faces?

Solution. There are eight faces, so four of them must be colored white, and the others black. To find the number of ways to do this, make the substitution, $z_i \to B^i + 1$ in equation 139 and find the coefficient of B^4. Making the substitution and simplifying, we get

$$p(B) = B^8 + B^7 + 3B^6 + 3B^5 + 7B^4 + 3B^3 + 3B^2 + B + 1$$

So the answer is 7. The number of ways to get 3, 2, and 1 black faces are 3, 3, and 1, respectively.

Problem 61. How many unique ways, under rotation, can you label the eight faces of an octahedron with the numbers 1 through 8 such that each number is used?

Solution. Let the eight numbers be symbolized by $x_1, x_2, \ldots x_8$. Then we make the substitution $x_i \to x_1^i + x_2^i + \cdots + x_8^i$ into the cycle index polynomial in equation 139 and find the coefficient of $x_1 x_2 \cdots x_8$. The only place where this term appears is in $z_1^8 \to (x_1 + x_2 + \cdots + x_8)^8$ with coefficient equal to 40320. Dividing this by 24, we get the answer 1680. There are 1680 unique ways to label the faces of an octahedron with all the numbers 1 through 8.

Problem 62. The dodecahedron is a Platonic solid with twelve faces, twenty vertices, and thirty edges. Each face is a regular pentagon. Figures 20, 21, and 22 show a dodecahedron looking into a face, vertex, and edge, respectively. The problem is to find the number of unique ways the faces of a dodecahedron can be colored with m colors. The colorings must be unique with respect to rotations.

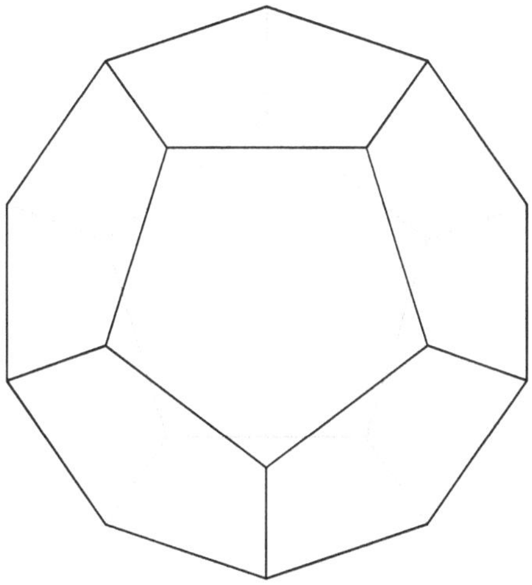

Figure 20: Dodecahedron face view.

Solution. As in the previous problems, we need to find how the faces are permuted under all possible rotations. There are six rotation axes through the centers of opposite faces, with possible rotations of 72°, 144°, 216°, and 288°. For each of these rotations, the two faces intersected by the axis remain fixed, and the faces adjacent to them are cyclically permuted. The cycle structure will then look like (1)(2)(3 4 5 6 7)(8 9 10 11 12) where faces 1 and 2 are fixed, and the two cycles of length five are the faces adjacent to them. The four

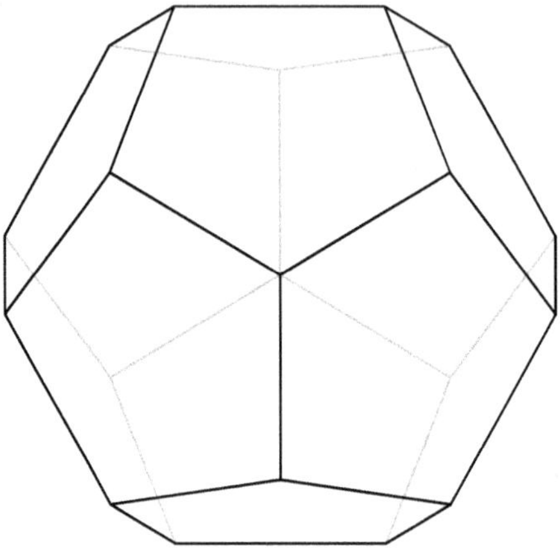

Figure 21: Dodecahedron vertex view.

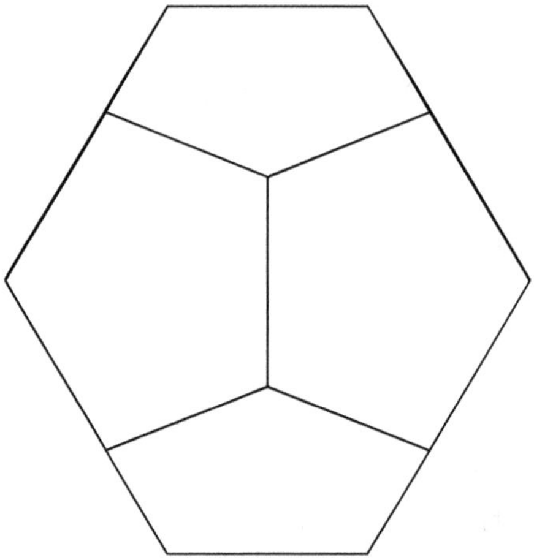

Figure 22: Dodecahedron edge view.

rotations each contribute $z_1^2 z_5^2$ to the cycle index, and there are six such rotation axes, so the total contribution is $24 z_1^2 z_5^2$.

The next set of rotation axes are through opposite pairs of vertices. The possible rotations are 120° and 240°. Three faces meet at each of the vertices, and they are cyclically permuted by the rotations. There are three faces that are adjacent to each pair of the faces adjacent to the rotation vertex. These faces are also cyclically permuted. So there are a total of four sets of three faces, cyclically permuted by the two rotations. The contribution to the cycle index for each rotation is z_3^4. There are ten rotation axes, each with two rotations, so the total contribution is $20 z_3^4$.

The final set of rotation axes are through the centers of opposite pairs of edges. The only possible rotation is 180° and it permutes all of the faces and pairs, so its contribution to the cycle index is z_2^6. There are 15 such rotation axes, so the total contribution is $15 z_2^6$. Including the identity, the dodecahedron has 60 rotational symmetries, and the cycle index polynomial is

$$p(z) = \frac{1}{60}(z_1^{12} + 15 z_2^6 + 20 z_3^4 + 24 z_1^2 z_5^2) \qquad (140)$$

To get the number of colorings, substitute $z_i \to m$.

$$p(m) = \frac{1}{60}(m^{12} + 15 m^6 + 44 m^4)$$

For the first 5 values of m, we have

m	1	2	3	4	5
$p(m)$	1	96	9099	280832	4073375

The integer sequence is A000545 in the OEIS.

Problem 63. How many unique ways, under rotation, can you color the faces of a dodecahedron black and white such that there are $k = 0, 1, 2, \ldots 6$ black faces?

Solution. In general, to find the number of ways to get k black faces, make the substitution, $z_i \to B^i + 1$ in equation 140 and find the coefficient of B^k. Making the substitution and simplifying, we get

$$\begin{aligned} p(B) &= B^{12} + B^{11} + 3B^{10} + 5B^9 + 12B^8 + 14B^7 + \\ &\quad 24B^6 + 14B^5 + 12B^4 + 5B^3 + 3B^2 + B + 1 \end{aligned}$$

So the number of ways to get $0, 1, 2, \ldots 6$ black faces are $1, 1, 3, 5, 12, 14, 24$, respectively.

Problem 64. How many unique ways, under rotation, can you label the twelve faces of a dodecahedron with the numbers 1 through 12 such that each number is used?

Solution. Let the twelve numbers be symbolized by $x_1, x_2, \ldots x_{12}$. Then we make the substitution $x_i \to x_1^i + x_2^i + \cdots + x_{12}^i$ into the cycle index polynomial in equation 140 and find the coefficient of $x_1 x_2 \cdots x_{12}$. The only place where this term appears is in $z_1^{12} \to$

$(x_1+x_2+\cdots+x_{12})^{12}$ with coefficient equal to 479001600. Dividing this by 60, we get the answer 7983360.

Problem 65. The Platonic solid with the most faces is the icosahedron. There are twenty faces, twelve vertices, and thirty edges. Each face is an equilateral triangle. Figures 23, 24, and 25 show an icosahedron looking into a face, vertex, and edge, respectively. The problem is to find the number of unique ways the faces of an icosahedron can be colored with m colors. The colorings must be unique with respect to rotations.

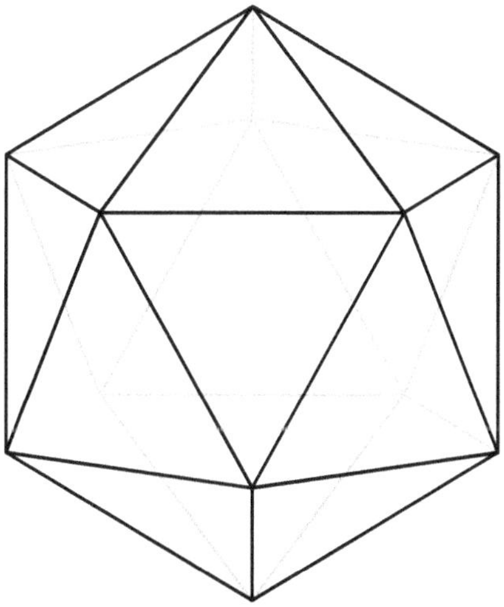

Figure 23: Icosahedron face view.

Solution. As in the previous problems, we need to find how the faces are permuted under all possible ro-

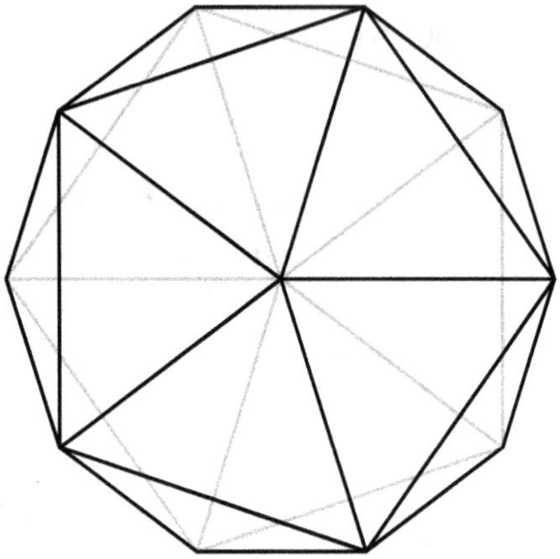

Figure 24: Icosahedron vertex view.

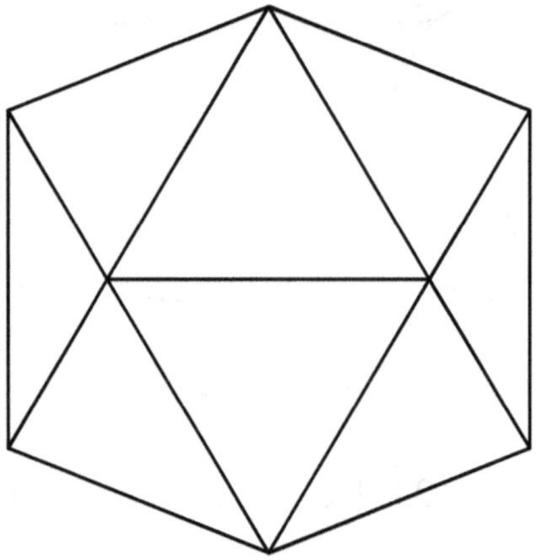

Figure 25: Icosahedron edge view.

tations. There are ten rotation axes through the centers of opposite faces, with possible rotations of 120° and 240°. These rotations cyclically permute the faces in groups of three. There are six such groups and the two faces intersected by the axes remain fixed, so the cycle index for these rotations is $20z_1^2z_3^6$. There are six rotation axes through opposite pairs of vertices, with possible rotations of 72°, 144°, 216°, and 288°. These rotations cyclically permute the faces in groups of five. There are four such groups, so the cycle index for these rotations is $24z_5^4$. There are 15 rotation axes through the centers of opposite pairs of edges with a possible rotation of 180°. These rotations permute pairs of faces. There are 10 such pairs, so the cycle index for these rotations is $15z_2^{10}$. Including the identity for a total of 60 rotational symmetries, we get the following cycle index polynomial

$$p(z) = \frac{1}{60}(z_1^{20} + 15z_2^{10} + 20z_1^2z_3^6 + 24z_5^4) \tag{141}$$

To get the number of colorings, substitute $z_i \to m$.

$$p(m) = \frac{1}{60}(m^{20} + 15m^{10} + 20m^8 + 24m^4)$$

For the first 4 values of m, we have

m	1	2	3	4
$p(m)$	1	17824	58130055	18325477888

The integer sequence is A054472 in the OEIS.

Problem 66. How many unique ways, under rotation, can you color the faces of an icosahedron black and white such that there are $k = 0, 1, 2, \ldots 10$ black faces?

Solution. In general, to find the number of ways to get k black faces, make the substitution, $z_i \to B^i + 1$ in equation 141 and find the coefficient of B^k. Making the substitution and simplifying, we get

$$\begin{aligned} p(B) &= B^{20} + B^{19} + 6B^{18} + 21B^{17} + 96B^{16} \\ &\quad +262B^{15} + 681B^{14} + 1302B^{13} + 2157B^{12} \\ &\quad +2806B^{11} + 3158B^{10} + 2806B^{9} + 2157B^{8} \\ &\quad +1302B^{7} + 681B^{6} + 262B^{5} + 96B^{4} + 21B^{3} \\ &\quad +6B^{2} + B + 1 \end{aligned}$$

So the number of ways to get $0, 1, 2, \ldots 10$ black faces are $1, 1, 6, 21, 96, 262, 681, 1302, 2157, 2806, 3158$, respectively.

Problem 67. How many unique ways, under rotation, can you label the twenty faces of an icosahedron with the numbers 1 through 20 such that each number is used?

Solution. Let the twenty numbers be symbolized by $x_1, x_2, \ldots x_{20}$. Then we make the substitution $x_i \to x_1^i + x_2^i + \cdots + x_{20}^i$ into the cycle index polynomial in equation 141 and find the coefficient of $x_1 x_2 \cdots x_{20}$. The only place where this term appears is in $z_1^{20} \to (x_1 + x_2 + \cdots + x_{20})^{20}$ with coefficient equal to

2432902008176640000.
Dividing this by 60, we get the answer
40548366802944000.

Problem 68. Find the number of words of length n composed of the digits $\{0, 1\}$ given that two words are considered equivalent if one is gotten from the other by changing 0 to 1, and 1 to 0. Find all words of length $n = 1, 2, 3, 4, 5$.

Solution. Every binary number of length n has one unique complement found by changing its 0's to 1's and its 1's to 0's. The complement of the complement is the original binary number. So if a number and its complement are considered equal, then the number of unique words is $2^n/2 = 2^{n-1}$.

Now let's see how this problem can be solved using a generalization of the Polya method. The permutation group for the letters is the symmetric group S_2 with cycle index $p_H(z_1, z_2) = \frac{1}{2}(z_1^2 + z_2)$. The permutation group for the word itself is just the identity with cycle index $p_G(z_1) = z_1^n$.

In general for these kinds of problems, we make the following substitution in the word cycle index

$$z_i^n \to \frac{\partial^n}{\partial z_i^n}$$

and we make the following substitution in the symbol

cycle index

$$\begin{aligned} z_i &\rightarrow e^{i\Sigma_i} \\ \Sigma_i &= \sum_{k=1}^{\infty} z_{i\cdot k} \end{aligned}$$

then we operate on p_H with p_G and evaluate the result at $z_i = 0$ for all i. For this problem we get

$$\begin{aligned} p_G \circ p_H &= \frac{\partial^n}{\partial z_1^n} \frac{1}{2} \left(e^{2\Sigma_1} + e^{2\Sigma_2} \right) \\ &= \frac{2^n}{2} \\ &= 2^{n-1} \end{aligned}$$

Examples for words of length $n = 1, 2, 3, 4, 5$ are shown below.

$$\begin{aligned} 1 &: 0 \\ 2 &: 00, 01 \\ 3 &: 000, 001, 010, 011 \\ 4 &: 0000, 0001, 0010, 0011, 0100, 0101, 0110, 0111 \\ 5 &: 00000, 00001, 00010, 00011, 00100, 00101, 00110, \\ &\quad 00111, 01000, 01001, 01010, 01011, 01100, 01101, \\ &\quad 01110, 01111 \end{aligned}$$

The examples above are generated by the following command:

```
autogen is2.aut k S 0 1
```

where `k` $= 1, 2, 3, 4, 5$, and `is2.aut` defines the automaton, consisting of the following lines:

```
3
S 00
0 00 11
1 01 11
```

The automaton is shown below.

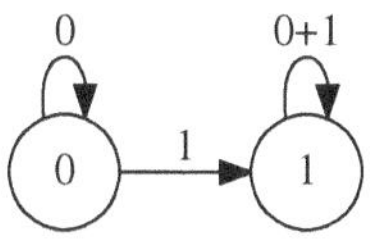

Problem 69. Repeat the previous problem for words composed of the digits $\{0, 1, 2\}$ given that two words are considered equivalent if one is gotten from the other under any permutation of the digit values. Find all words of length $n = 1, 2, 3, 4, 5$.

Solution. Once again, the permutation group for the word itself is just the identity with cycle index $p_G(z_1) = z_1^n$. This time the permutation group for the letters is the symmetric group S_3 with cycle index

$$p_H(z_1, z_2, z_3) = \frac{1}{6}(z_1^3 + 3z_1z_2 + 2z_3)$$

Making the substitutions and evaluating the result as in the previous problem, we have

$$\begin{aligned} p_G \circ p_H &= \frac{\partial^n}{\partial z_1^n} \frac{1}{6} \left(e^{3\Sigma_1} + 3e^{\Sigma_1 + 2\Sigma_2} + 2e^{3\Sigma_3} \right) \\ &= \frac{1}{6}(3^n + 3) \end{aligned}$$

The first few values are

n	1	2	3	4	5	6	7	8	9	10
$a(n)$	1	2	5	14	41	122	365	1094	3281	9842

The integer sequence is A007051 in the OEIS. Examples for words of length $n = 1, 2, 3, 4, 5$ are shown below.

$$\begin{aligned}
1 &: 0 \\
2 &: 00, 01 \\
3 &: 000, 001, 010, 011, 012 \\
4 &: 0000, 0001, 0010, 0011, 0012, 0100, 0101, 0102, \\
&\quad 0110, 0111, 0112, 0120, 0121, 0122 \\
5 &: 00000, 00001, 00010, 00011, 00012, 00100, 00101, \\
&\quad 00102, 00110, 00111, 00112, 00120, 00121, 00122, \\
&\quad 01000, 01001, 01002, 01010, 01011, 01012, 01020, \\
&\quad 01021, 01022, 01100, 01101, 01102, 01110, 01111, \\
&\quad 01112, 01120, 01121, 01122, 01200, 01201, 01202, \\
&\quad 01210, 01211, 01212, 01220, 01221, 01222
\end{aligned}$$

The examples above are generated by the following command:

```
autogen is3.aut k S 0 1 2
```

where k $= 1, 2, 3, 4, 5$, and `is3.aut` defines the automaton, consisting of the following lines:

```
4
S 00
0 00 11
1 01 11 22
2 02 12 22
```

The automaton is shown below.

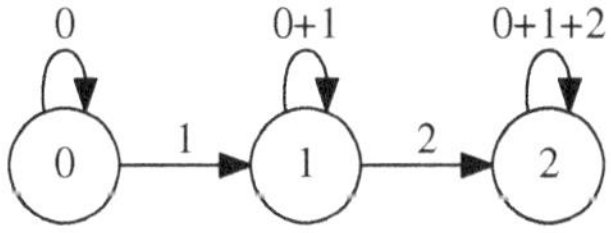

Problem 70. Repeat the previous problem for words composed of the digits $\{0, 1, 2, 3\}$ given that two words are considered equivalent if one is gotten from the other under any permutation of the digit values. Find all words of length $n = 1, 2, 3, 4, 5, 6$.

Solution. The permutation group for the word is the identity with cycle index $p_G(z_1) = z_1^n$. The permutation group for the letters is the symmetric group S_4 with cycle index

$$p_H(z_1, z_2, z_3, z_4) = \frac{1}{24}(z_1^4 + 6z_1^2 z_2 + 8z_1 z_3 + 3z_2^2 + 6z_4)$$

Making the substitutions and evaluating the result as in the previous problem, we have (ignoring terms that do not contain z_1)

$$\begin{aligned} p_G \circ p_H &= \frac{\partial^n}{\partial z_1^n} \frac{1}{24} \left(e^{4\Sigma_1} + 6e^{2\Sigma_1 + 2\Sigma_2} + 8e^{\Sigma_1 + 3\Sigma_3}\right) \\ &= \frac{1}{24}(4^n + 6 \cdot 2^n + 8) \end{aligned}$$

The first few values are

n	1	2	3	4	5	6	7	8
$a(n)$	1	2	5	15	51	187	715	2795

n	9	10
$a(n)$	11051	43947

The integer sequence is A007581 in the OEIS. Examples for words of length $n = 1, 2, 3, 4, 5, 6$ are the same as those in the previous problem, plus the ones shown below.

4 : 0123

5 : 00123, 01023, 01123, 01203, 01213, 01223, 01230, 01231, 01232, 01233

6 : 000000, 000001, 000010, 000011, 000012, 000100, 000101, 000102, 000110, 000111, 000112, 000120, 000121, 000122, 000123, 001000, 001001, 001002, 001010, 001011, 001012, 001020, 001021, 001022, 001023, 001100, 001101, 001102, 001110, 001111, 001112, 001120, 001121, 001122, 001123, 001200, 001201, 001202, 001203, 001210, 001211, 001212, 001213, 001220, 001221, 001222, 001223, 001230, 001231, 001232, 001233, 010000, 010001, 010002, 010010, 010011, 010012, 010020, 010021, 010022, 010023, 010100, 010101, 010102, 010110, 010111, 010112, 010120, 010121, 010122, 010123, 010200, 010201, 010202, 010203, 010210, 010211, 010212, 010213, 010220, 010221, 010222, 010223, 010230, 010231, 010232, 010233, 011000, 011001, 011002, 011010, 011011, 011012, 011020, 011021, 011022, 011023, 011100, 011101, 011102, 011110, 011111, 011112, 011120, 011121, 011122, 011123, 011200, 011201, 011202, 011203, 011210, 011211, 011212, 011213, 011220, 011221, 011222, 011223, 011230, 011231, 011232, 011233, 012000, 012001, 012002, 012003, 012010, 012011, 012012, 012013, 012020, 012021, 012022, 012023, 012030, 012031, 012032, 012033, 012100, 012101, 012102, 012103, 012110, 012111, 012112, 012113, 012120, 012121, 012122, 012123, 012130, 012131, 012132, 012133, 012200, 012201, 012202, 012203, 012210, 012211, 012212, 012213, 012220, 012221, 012222, 012223, 012230, 012231, 012232, 012233, 012300, 012301, 012302, 012303, 012310, 012311, 012312, 012313, 012320, 012321, 012322, 012323, 012330, 012331, 012332, 012333

The examples above are generated by the following command:

```
autogen is4.aut k S 0 1 2 3
```

where `k` $= 1, 2, 3, 4, 5, 6$, and `is4.aut` defines the automaton, consisting of the following lines:

```
5
S 00
0 00 11
1 01 11 22
2 02 12 22 33
3 03 13 23 33
```

The automaton is shown below.

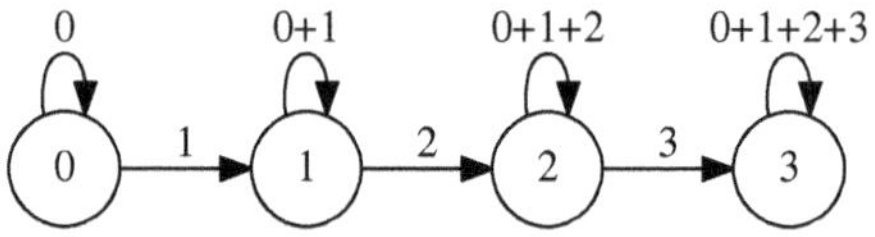

Problem 71. Repeat the previous problem for words composed of the digits $\{0, 1, 2, 3, 4\}$ given that two words are considered equivalent if one is gotten from the other under any permutation of the digit values. Find all words of length $n = 1, 2, 3, 4, 5, 6$.

Solution. The permutation group for the word is the identity with cycle index $p_G(z_1) = z_1^n$. The permutation group for the letters is the symmetric group S_5 with cycle index

$$\begin{aligned} p_H(z) &= \tfrac{1}{120}(z_1^5 + 10z_1^3 z_2 + 20z_1^2 z_3 + 15z_1 z_2^2 + 30z_1 z_4 \\ &\qquad +20z_2 z_3 + 24z_5) \end{aligned}$$

Making the substitutions and evaluating the result as in the previous problem, we have (ignoring terms that do not contain z_1)

$$\begin{aligned} p_G \circ p_H &= \frac{\partial^n}{\partial z_1^n} \frac{1}{120} \left(e^{5\Sigma_1} + 10e^{3\Sigma_1} + 20e^{2\Sigma_1} + 45e^{\Sigma_1}\right) \\ &= \frac{1}{120}(5^n + 10 \cdot 3^n + 20 \cdot 2^n + 45) \end{aligned}$$

The first few values are

n	1	2	3	4	5	6	7	8	9
$a(n)$	1	2	5	15	52	202	855	3845	18002

The integer sequence is A056272 in the OEIS. Examples for words of length $n = 1, 2, 3, 4, 5, 6$ are the same as those in the previous problem, plus the ones shown below.

5 : 01234

6 : 001234, 010234, 011234, 012034, 012134, 012234, 012304, 012314, 012324, 012334, 012340, 012341, 012342, 012343, 012344

The examples above are generated by the following command:

```
autogen is5.aut k S 0 1 2 3 4
```

where `k` $= 1, 2, 3, 4, 5, 6$, and `is5.aut` defines the automaton, consisting of the following lines:

```
6
S 00
0 00 11
1 01 11 22
2 02 12 22 33
3 03 13 23 33 44
4 04 14 24 34 44
```

The automaton is shown below.

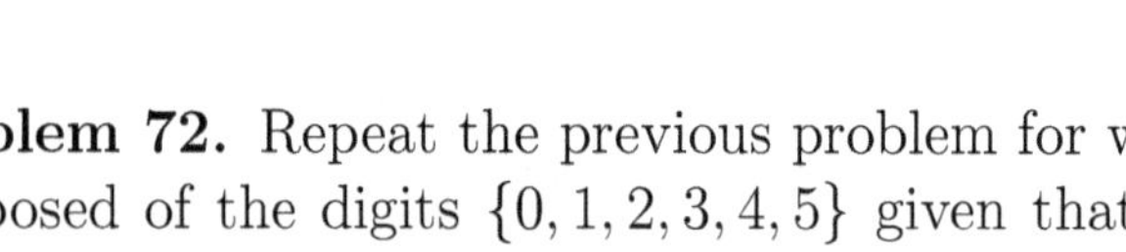

Problem 72. Repeat the previous problem for words composed of the digits $\{0, 1, 2, 3, 4, 5\}$ given that two words are considered equivalent if one is gotten from the other under any permutation of the digit values.

Solution. The permutation group for the word is the identity with cycle index $p_G(z_1) = z_1^n$. The permutation group for the letters is the symmetric group S_6. This is a large group, and we don't need all the terms in the cycle index. Since we are taking derivatives with respect to z_1, only the terms that contain z_1 are needed. Those terms are: z_1^6, $z_1^4 z_2$, $z_1^3 z_3$, $z_1^2 z_4$, $z_1^2 z_2^2$, $z_1 z_5$, $z_1 z_2 z_3$.

We can calculate the coefficients of these terms using simple combinatorial reasoning. There is only one way to have 6 one cycles, so the coefficient of z_1^6 is 1. For $z_1^4 z_2$ there are $\binom{6}{4} = 15$ ways to have 4 one cycles, and each way uniquely determines the two cycle, so the coefficient is 15. For $z_1^3 z_3$ there are $\binom{6}{3} = 20$ ways to have 3 one cycles, and for each of these there are two possible three cycles, so the coefficient is 40. For $z_1^2 z_4$ there are $\binom{6}{2} = 15$ double one cycles, and for each there are $3! = 6$ possible four cycles[7], so the coefficient is $15 \cdot 6 = 90$. For $z_1^2 z_2^2$ there are $\binom{6}{2} = 15$ double one cycles, and from the remaining four elements, 2 two cycles can be constructed in $\frac{1}{2}\binom{4}{2} = 3$ ways (remember $(3\ 4)(5\ 6)$ is the same as $(5\ 6)(3\ 4)$ hence the factor of $\frac{1}{2}$) so the coefficient is $15 \cdot 3 = 45$. For $z_1 z_5$ there are 6 one cycles, and $4! = 24$ five cycles for each, so the coefficient is $6 \cdot 24 = 144$. For $z_1 z_2 z_3$ there are 6 one cycles, $\binom{5}{2} = 10$ two cycles, and $2! = 2$ three cycles, so the coefficient is $6 \cdot 10 \cdot 2 = 120$.

[7]Note that in general, the number of cycles of length k is $(k-1)!$ since they can all be gotten by fixing one of the elements and using all $(k-1)!$ permutations of the other elements.

The relevant part of the cycle index for S_6 is then

$$\begin{aligned} p_H(z) &= \tfrac{1}{720}(z_1^6 + 15z_1^4 z_2 + 40z_1^3 z_3 + 90z_1^2 z_4 + \\ &\quad 45z_1^2 z_2^2 + 144z_1 z_5 + 120z_1 z_2 z_3) \end{aligned}$$

We leave it to the reader to evaluate $p_G \circ p_H$ as in the previous problems, and show that the number of words of length n is given by

$$a(n) = \frac{1}{720}(6^n + 15 \cdot 4^n + 40 \cdot 3^n + 135 \cdot 2^n + 264)$$

The first few values are

n	1	2	3	4	5	6	7	8
$a(n)$	1	2	5	15	52	203	876	4111

n	9	10
$a(n)$	20648	109299

The integer sequence is A056273 in the OEIS.

Problem 73. Note that the first six numbers in the sequence of word counts for the last problem:
$1, 2, 5, 15, 52, 203$
are equal to the first six Bell numbers, B_n for $n = 1, 2, 3, \ldots 6$. The Bell numbers are sequence A000110 in the OEIS. The nth Bell number is equal to the number of ways a set of n elements can be partitioned into disjoint subsets whose union gives the set. For the set $\{1, 2, 3\}$ for example, there are 5 possible partitions. They are
$\{1, 2, 3\}$, $\{1\}\{2, 3\}$, $\{2\}\{1, 3\}$, $\{3\}\{1, 2\}$, $\{1\}\{2\}\{3\}$

so $B_3 = 5$. Show that this is not a coincidence, and that there is a connection between the Bell numbers and the word counts that were calculated in the previous few problems.

Solution. The unique words of length 3 from the previous few problems are

$$\begin{array}{ccccc} 000 & 001 & 010 & 011 & 012 \\ 123 & 123 & 123 & 123 & 123 \end{array}$$

Under each symbol we have written its position number in the word. Assume that the position numbers are actually elements of the set $\{1, 2, 3\}$ and the symbol values indicate subsets into which the elements are placed. All elements with the same value are placed in the same subset. The word 000 for example, indicates that elements 1, 2, and 3 all go into the same subset. Word 001 puts 1 and 2 in the same subset, and 3 in a set by itself, producing the partition $\{3\}\{1, 2\}$. Likewise, 010 produces the partition $\{2\}\{1, 3\}$, 011 produces the partition $\{1\}\{2, 3\}$, and 012 produces the partition $\{1\}\{2\}\{3\}$. Each word produces a partition, and permuting the symbol values in a word does not change the partition. For example, changing 011 to 100 still produces the partition $\{1\}\{2, 3\}$. When using an alphabet of m symbols, there will be a one to one correspondence between words of length n and partitions of a set of n elements as long as $n \leq m$. For $n > m$, the words will represent partitions into at most m subsets.

Problem 74. Find an automaton that can generate all the partitions of a set of n elements into at most four subsets, and use the automaton to find a generating function for the number of such partitions.

Solution. Choose one of the elements of the set and place it in a set by itself. Call this set 0. The next element can either go into set 0 or a new set called set 1. Suppose we put it in set 1, then the next element can go in set 0, set 1, or a new set called set 2. The process we just described can be represented by the automaton shown below.

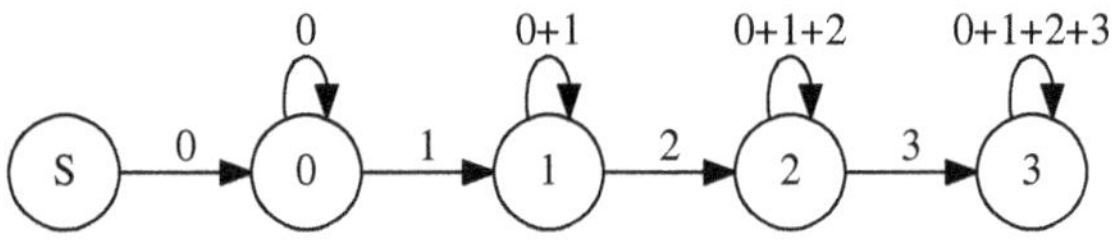

All partitions into at most four subsets can be generated by starting at state S and ending in any of the other states. The adjacency matrix for the automaton is

$$\mathbf{A} = \begin{pmatrix} 0 & 1 & 0 & 0 & 0 \\ 0 & 1 & 1 & 0 & 0 \\ 0 & 0 & 2 & 1 & 0 \\ 0 & 0 & 0 & 3 & 1 \\ 0 & 0 & 0 & 0 & 4 \end{pmatrix} \qquad (142)$$

Calculating $(\mathbf{I} - z\mathbf{A})^{-1}$ and summing the first row gives us the generating function.

$$G(z) = \frac{3z^3 - 9z^2 + 6z - 1}{(z-1)(2z-1)(4z-1)} \tag{143}$$

The Taylor series expansion is

$$G(z) = 1+z+2z^2+5z^3+15z^4+51z^5+187z^6+715z^7+2795z^8 \ldots \tag{144}$$

The integer sequence is identical to the sequence we calculated in Problem 68.

Problem 75. Generalize the previous problem to find the generating function for the number of partitions of a set of n elements into at most m subsets.

Solution. Generalizing this problem is most easily done in terms of the regular expression for the automaton. We'll start by looking at the regular expression for the automaton in the previous problem. We can end in the states $0, 1, 2, 3$ and there are no backward paths to previous states, so the regular expression can easily be written down by inspection.

$$\begin{aligned} R &= 00^* + 00^*1(0+1)^* + 00^*1(0+1)^*2(0+1+2)^* + \\ &\quad 00^*1(0+1)^*2(0+1+2)^*3(0+1+2+3)^* \end{aligned}$$

And converting this into a generating function using the standard technique gives

$$G(z) = \frac{z}{1-z} + \frac{z^2}{(1-z)(1-2z)} + \frac{z^3}{(1-z)(1-2z)(1-3z)} + \frac{z^4}{(1-z)(1-2z)(1-3z)(1-4z)} \tag{145}$$

The generalization to the case of m subsets should then be obvious. The generating function for the case of m subsets is

$$G(z) = \sum_{k=1}^{m} \prod_{j=1}^{k} \frac{z}{1-jz} \tag{146}$$

Problem 76. Find the number of words of length n composed of the digits $\{0, 1\}$ given that two words are considered equivalent if one is gotten from the other by changing 0 to 1, and 1 to 0, and/or if one is the reverse of the other.

Solution. The permutation group for the word itself is

$$p_G(z) = \frac{1}{2}\begin{cases} z_1^n + z_2^{n/2}, & n = 0, 2, 4, \ldots \\ z_1^n + z_1 z_2^{(n-1)/2}, & n = 1, 3, 5, \ldots \end{cases} \tag{147}$$

The permutation group for the letters is the symmetric group S_2 with cycle index

$$p_H(z_1, z_2) = \frac{1}{2}(z_1^2 + z_2)$$

Making the substitutions as in the previous problem, we have for even n

$$p_G \circ p_H = \frac{1}{2}\left(\frac{\partial^n}{\partial z_1^n} + \frac{\partial^{n/2}}{\partial z_2^{n/2}}\right)\frac{1}{2}\left(e^{2\Sigma_1} + e^{2\Sigma_2}\right)$$

and for odd n

$$\begin{aligned} p_G \circ p_H &= \frac{1}{2}\left(\frac{\partial^n}{\partial z_1^n} + \frac{\partial}{\partial z_1}\frac{\partial^{(n-1)/2}}{\partial z_2^{(n-1)/2}}\right) \\ &\quad \frac{1}{2}\left(e^{2\Sigma_1} + e^{2\Sigma_2}\right) \end{aligned}$$

Evaluating these expressions, we get the following formula for the number of words

$$a(n) = \frac{1}{4}\begin{cases} 2^n + 2 \cdot 2^{n/2}, & n = 2, 4, 6 \ldots \\ 2^n + 2 \cdot 2^{(n-1)/2}, & n = 1, 3, 5 \ldots \end{cases} \tag{148}$$

The first few values are

n	1	2	3	4	5	6	7	8	9	10
$a(n)$	1	2	3	6	10	20	36	72	136	272

The integer sequence is A005418 in the OEIS. Examples for words of length $n = 1, 2, 3, 4, 5$ are shown below.

$$\begin{aligned}
1 &: 0 \\
2 &: 00, 01 \\
3 &: 000, 001, 010 \\
4 &: 0000, 0001, 0010, 0011, 0101, 0110 \\
5 &: 00000, 00001, 00010, 00011, 00100, 00101, \\
&\quad 00110, 01001, 01010, 01110
\end{aligned}$$

Problem 77. Repeat the previous problem for words composed of the digits $\{0, 1, 2\}$ given that two words are considered equivalent if one is gotten from the other under any permutation of the digit values, and/or if one is the reverse of the other.

Solution. The permutation group for the word itself is the same as in the previous problem. The permutation group for the letters is the symmetric group S_3 with cycle index

$$p_H(z_1, z_2, z_3) = \frac{1}{6}(z_1^3 + 3z_1 z_2 + 2z_3)$$

Making the substitutions as in the previous problem, we have for even n

$$p_G \circ p_H = \frac{1}{12}\left(\frac{\partial^n}{\partial z_1^n} + \frac{\partial^{n/2}}{\partial z_2^{n/2}}\right)\left(e^{3\Sigma_1} + 3e^{\Sigma_1}e^{2\Sigma_2} + 2e^{3\Sigma_3}\right)$$

and for odd n

$$p_G \circ p_H = \frac{1}{12}\left(\frac{\partial^n}{\partial z_1^n} + \frac{\partial}{\partial z_1}\frac{\partial^{(n-1)/2}}{\partial z_2^{(n-1)/2}}\right) \circ \left(e^{3\Sigma_1} + 3e^{\Sigma_1}e^{2\Sigma_2} + 2e^{3\Sigma_3}\right)$$

Evaluating these expressions, we get the following formula for the number of words

$$a(n) = \frac{1}{12}\begin{cases} 3^n + 4 \cdot 3^{n/2} + 3, & n = 2, 4, 6 \ldots \\ 3^n + 2 \cdot 3^{(n+1)/2} + 3, & n = 1, 3, 5 \ldots \end{cases} \quad (149)$$

The first few values are

n	1	2	3	4	5	6	7	8	9	10
$a(n)$	1	2	4	10	25	70	196	574	1681	5002

The integer sequence is A001998 in the OEIS.

Problem 78. Repeat the previous problem for words composed of the digits $\{0, 1, 2, 3\}$ given that two words are considered equivalent if one is gotten from the other under any permutation of the digit values, and/or if one is the reverse of the other.

Solution. The permutation group for the word itself is the same as in the previous problem. The permutation group for the letters is the symmetric group S_4 with cycle index

$$p_H(z_1, z_2, z_3, z_4) = \frac{1}{24}(z_1^4 + 6z_1^2 z_2 + 8z_1 z_3 + 3z_2^2 + 6z_4)$$

Making the substitutions as in the previous two problems, and evaluating the results, we get the following formula for the number of words

$$a(n) = \frac{1}{48}\begin{cases} 4^n + 10 \cdot 4^{n/2} + 6 \cdot 2^n + 16, & n = 2, 4, 6 \ldots \\ 4^n + 16 \cdot 4^{(n-1)/2} + 6 \cdot 2^n + 16, & n = 1, 3, 5 \ldots \end{cases} \tag{150}$$

The first few values are

n	1	2	3	4	5	6	7	8
$a(n)$	1	2	4	11	31	107	379	1451

n	9	10
$a(n)$	5611	22187

The integer sequence is A056323 in the OEIS.

Problem 79. Find the number of necklaces of length n composed of two colors where two necklaces are considered equivalent if one is gotten from the other by any rotation and/or exchange of the two colors.

Solution. The permutation group for the necklace is

$$p_G(z) = \frac{1}{n}\sum_{d|n} \phi(d) z_d^{n/d} \tag{151}$$

The permutation group for the letters is the symmetric group S_2 with cycle index

$$p_H(z_1, z_2) = \frac{1}{2}(z_1^2 + z_2)$$

Making the substitutions as in the previous problems give us

$$p_G \circ p_H = \frac{1}{2n} \sum_{d|n} \phi(d) \frac{\partial^{n/d}}{\partial z_d^{n/d}} \left(e^{2\Sigma_1} + e^{2\Sigma_2}\right)$$

The derivatives of the two terms evaluated at $z_i = 0$ are

$$\frac{\partial^{n/d}}{\partial z_d^{n/d}} e^{2\Sigma_1} = 2^{n/d}$$

$$\frac{\partial^{n/d}}{\partial z_d^{n/d}} e^{2\Sigma_2} = \begin{cases} 2^{n/d}, & d = 2, 4, 6 \ldots \\ 0, & d = 1, 3, 5 \ldots \end{cases}$$

The number of necklaces is then given by

$$a(n) = \frac{1}{2n} \sum_{d|n} \phi(d) 2^{n/d} \begin{cases} 2, & d = 2, 4, 6 \ldots \\ 1, & d = 1, 3, 5 \ldots \end{cases} \tag{152}$$

The first few values are

n	1	2	3	4	5	6	7	8	9	10
$a(n)$	1	2	2	4	4	8	10	20	30	56

The integer sequence is A000013 in the OEIS. Examples for words of length $n = 1, 2, 3, 4, 5, 6$ are shown below.

$$\begin{aligned}
1 &: 0 \\
2 &: 00, 01 \\
3 &: 000, 001 \\
4 &: 0000, 0001, 0011, 0101 \\
5 &: 00000, 00001, 00011, 00101 \\
6 &: 000000, 000001, 000011, 000101, \\
&\quad 000111, 001001, 001011, 010101
\end{aligned}$$

Problem 80. Find the number of necklaces of length n composed of three colors where two necklaces are considered equivalent if one is gotten from the other by any rotation and/or permutation of the three colors.

Solution. The permutation group for the necklace is

$$p_G(z) = \frac{1}{n} \sum_{d|n} \phi(d) z_d^{n/d} \tag{153}$$

The permutation group for the letters is the symmetric group S_3 with cycle index

$$p_H(z_1, z_2, z_3) = \frac{1}{6}(z_1^3 + 3z_1 z_2 + 2z_3)$$

Making the substitutions as in the previous problems give us

$$p_G \circ p_H = \frac{1}{2n} \sum_{d|n} \phi(d) \frac{\partial^{n/d}}{\partial z_d^{n/d}} \left(e^{3\Sigma_1} + 3e^{\Sigma_1 + 2\Sigma_2} + 2e^{3\Sigma_3} \right)$$

The derivatives of the three terms evaluated at $z_i = 0$ are

$$\frac{\partial^{n/d}}{\partial z_d^{n/d}} e^{3\Sigma_1} = 3^{n/d}$$

$$\frac{\partial^{n/d}}{\partial z_d^{n/d}} e^{\Sigma_1 + 2\Sigma_2} = \begin{cases} 3^{n/d}, & d = 2, 4, 6 \ldots \\ 1, & d = 1, 3, 5 \ldots \end{cases}$$

$$\frac{\partial^{n/d}}{\partial z_d^{n/d}} e^{3\Sigma_3} = \begin{cases} 3^{n/d}, & d \mod 3 = 0 \\ 0, & d \mod 3 \neq 0 \end{cases}$$

The number of necklaces is then given by

$$a(n) = \frac{1}{6n} \sum_{d|n} \phi(d) \begin{cases} 6 \cdot 3^{n/d}, & d = 6, 12, 18, 24, 30 \ldots \\ 4 \cdot 3^{n/d}, & d = 2, 4, 8, 10, 14 \ldots \\ 3 \cdot 3^{n/d} + 3, & d = 3, 9, 15, 21, 27 \ldots \\ 3^{n/d} + 3, & d = 1, 5, 7, 11, 13 \ldots \end{cases} \tag{154}$$

The first few values are

n	1	2	3	4	5	6	7	8	9	10
$a(n)$	1	2	3	6	9	26	53	146	369	1002

The integer sequence is A002076 in the OEIS.

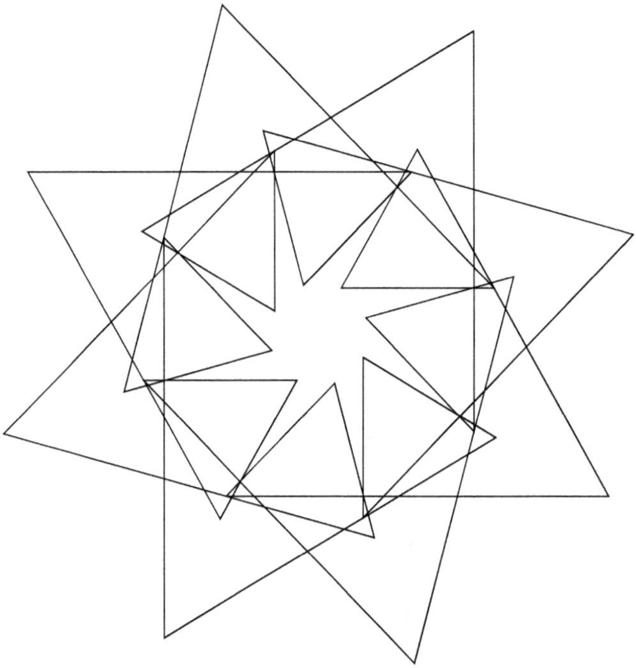

TOTIENT FUNCTION

The totient function of the integer n is equal to the number of integers from 1 to n that are relatively prime to n. In other words, the totient function counts the integers k, $1 \leq k \leq n$, for which the greatest common divisor of k and n is equal to 1, $\gcd(n,k) = 1$. The totient of n is usually written as $\phi(n)$ and is often called Euler's totient function, or Euler's phi function.

If $n = p$ is a prime, then all the integers from 1 to p are relatively prime to p except for p itself, therefore $\phi(p) = p - 1$. More generally, if $n = p^k$ for some $k > 0$ then the only integers from 1 to p^k that are not relatively prime to p^k are the p^{k-1} integers p, $2p$, $3p$, ..., $p^{k-1}p$, therefore $\phi(p^k) = p^k - p^{k-1}$.

In number theory, ϕ is called a multiplicative function, which means that if $\gcd(n,m) = 1$, then $\phi(n,m) = \phi(n)\phi(m)$. This, and the result above for $\phi(p^k)$ can be used to derived an interesting formula for $\phi(n)$. If the prime factorization of n is $n = p_1^{k_1} p_2^{k_2} \cdots p_m^{k_m}$ then

$$\begin{aligned} \phi(n) &= \prod_{i=1}^{m} \phi(p_i^{k_i}) \\ &= \prod_{i=1}^{m} (p_i^{k_i} - p_i^{k_i-1}) \\ &= \prod_{i=1}^{m} p_i^{k_i}(1 - 1/p_i) \\ &= n\prod_{i=1}^{m}(1 - 1/p_i) \qquad (155) \end{aligned}$$

Another useful property is that summing $\phi(d)$ over all the divisors, d, of n gives n.

$$\sum_{d|n} \phi(d) = n \qquad (156)$$

The following table lists values of $\phi(n)$ for $n = 1$ to 199.

	0	1	2	3	4	5	6	7	8	9
0		1	1	2	2	4	2	6	4	6
10	4	10	4	12	6	8	8	16	6	18
20	8	12	10	22	8	20	12	18	12	28
30	8	30	16	20	16	24	12	36	18	24
40	16	40	12	42	20	24	22	46	16	42
50	20	32	24	52	18	40	24	36	28	58
60	16	60	30	36	32	48	20	66	32	44
70	24	70	24	72	36	40	36	60	24	78
80	32	54	40	82	24	64	42	56	40	88
90	24	72	44	60	46	72	32	96	42	60
100	40	100	32	102	48	48	52	106	36	108
110	40	72	48	112	36	88	56	72	58	96
120	32	110	60	80	60	100	36	126	64	84
130	48	130	40	108	66	72	64	136	44	138
140	48	92	70	120	48	112	72	84	72	148
150	40	150	72	96	60	120	48	156	78	104
160	64	132	54	162	80	80	82	166	48	156
170	64	108	84	172	56	120	80	116	88	178
180	48	180	72	120	88	144	60	160	92	108
190	72	190	64	192	96	96	84	196	60	198

Divisors of integers from 1 to 199 are shown below.

1: 1

2: 1, 2

3: 1, 3

4: 1, 2, 4

5: 1, 5

6: 1, 2, 3, 6

7: 1, 7

8: 1, 2, 4, 8

9: 1, 3, 9

10: 1, 2, 5, 10

11: 1, 11

12: 1, 2, 3, 4, 6, 12

13: 1, 13

14: 1, 2, 7, 14

15: 1, 3, 5, 15

16: 1, 2, 4, 8, 16

17: 1, 17

18: 1, 2, 3, 6, 9, 18

19: 1, 19

20: 1, 2, 4, 5, 10, 20

21: 1, 3, 7, 21

22: 1, 2, 11, 22

23: 1, 23

24: 1, 2, 3, 4, 6, 8, 12, 24

25: 1, 5, 25

26: 1, 2, 13, 26

27: 1, 3, 9, 27

28: 1, 2, 4, 7, 14, 28

29: 1, 29

30: 1, 2, 3, 5, 6, 10, 15, 30

31: 1, 31

32: 1, 2, 4, 8, 16, 32

33: 1, 3, 11, 33

34: 1, 2, 17, 34

35: 1, 5, 7, 35

36: 1, 2, 3, 4, 6, 9, 12, 18, 36

37: 1, 37

38: 1, 2, 19, 38

39: 1, 3, 13, 39

40: 1, 2, 4, 5, 8, 10, 20, 40

41: 1, 41

42: 1, 2, 3, 6, 7, 14, 21, 42

43: 1, 43

44: 1, 2, 4, 11, 22, 44

45: 1, 3, 5, 9, 15, 45

46: 1, 2, 23, 46

47: 1, 47

48: 1, 2, 3, 4, 6, 8, 12, 16, 24, 48

49: 1, 7, 49

50: 1, 2, 5, 10, 25, 50

51: 1, 3, 17, 51

52: 1, 2, 4, 13, 26, 52

53: 1, 53

54: 1, 2, 3, 6, 9, 18, 27, 54

55: 1, 5, 11, 55

56: 1, 2, 4, 7, 8, 14, 28, 56

57: 1, 3, 19, 57

58: 1, 2, 29, 58

59: 1, 59

60: 1, 2, 3, 4, 5, 6, 10, 12, 15, 20, 30, 60

61: 1, 61

62: 1, 2, 31, 62

63: 1, 3, 7, 9, 21, 63

64: 1, 2, 4, 8, 16, 32, 64

65: 1, 5, 13, 65

66: 1, 2, 3, 6, 11, 22, 33, 66

67: 1, 67

68: 1, 2, 4, 17, 34, 68

69: 1, 3, 23, 69

70: 1, 2, 5, 7, 10, 14, 35, 70

71: 1, 71

72: 1, 2, 3, 4, 6, 8, 9, 12, 18, 24, 36, 72

73: 1, 73

74: 1, 2, 37, 74

75: 1, 3, 5, 15, 25, 75

76: 1, 2, 4, 19, 38, 76

77: 1, 7, 11, 77

78: 1, 2, 3, 6, 13, 26, 39, 78

79: 1, 79

80: 1, 2, 4, 5, 8, 10, 16, 20, 40, 80

81: 1, 3, 9, 27, 81

82: 1, 2, 41, 82

83: 1, 83

84: 1, 2, 3, 4, 6, 7, 12, 14, 21, 28, 42, 84

85: 1, 5, 17, 85

86: 1, 2, 43, 86

87: 1, 3, 29, 87

88: 1, 2, 4, 8, 11, 22, 44, 88

89: 1, 89

90: 1, 2, 3, 5, 6, 9, 10, 15, 18, 30, 45, 90

91: 1, 7, 13, 91

92: 1, 2, 4, 23, 46, 92

93: 1, 3, 31, 93

94: 1, 2, 47, 94

95: 1, 5, 19, 95

96: 1, 2, 3, 4, 6, 8, 12, 16, 24, 32, 48, 96

97: 1, 97

98: 1, 2, 7, 14, 49, 98

99: 1, 3, 9, 11, 33, 99

100: 1, 2, 4, 5, 10, 20, 25, 50, 100

101: 1, 101

102: 1, 2, 3, 6, 17, 34, 51, 102

103: 1, 103

104: 1, 2, 4, 8, 13, 26, 52, 104

105: 1, 3, 5, 7, 15, 21, 35, 105

106: 1, 2, 53, 106

107: 1, 107

108: 1, 2, 3, 4, 6, 9, 12, 18, 27, 36, 54, 108

109: 1, 109

110: 1, 2, 5, 10, 11, 22, 55, 110

111: 1, 3, 37, 111

112: 1, 2, 4, 7, 8, 14, 16, 28, 56, 112

113: 1, 113

114: 1, 2, 3, 6, 19, 38, 57, 114

115: 1, 5, 23, 115

116: 1, 2, 4, 29, 58, 116

117: 1, 3, 9, 13, 39, 117

118: 1, 2, 59, 118

119: 1, 7, 17, 119

120: 1, 2, 3, 4, 5, 6, 8, 10, 12, 15, 20, 24, 30, 40, 60, 120

121: 1, 11, 121

122: 1, 2, 61, 122

123: 1, 3, 41, 123

124: 1, 2, 4, 31, 62, 124

125: 1, 5, 25, 125

126: 1, 2, 3, 6, 7, 9, 14, 18, 21, 42, 63, 126

127: 1, 127

128: 1, 2, 4, 8, 16, 32, 64, 128

129: 1, 3, 43, 129

130: 1, 2, 5, 10, 13, 26, 65, 130

131: 1, 131

132: 1, 2, 3, 4, 6, 11, 12, 22, 33, 44, 66, 132

133: 1, 7, 19, 133

134: 1, 2, 67, 134

135: 1, 3, 5, 9, 15, 27, 45, 135

136: 1, 2, 4, 8, 17, 34, 68, 136

137: 1, 137

138: 1, 2, 3, 6, 23, 46, 69, 138

139: 1, 139

140: 1, 2, 4, 5, 7, 10, 14, 20, 28, 35, 70, 140

141: 1, 3, 47, 141

142: 1, 2, 71, 142

143: 1, 11, 13, 143

144: 1, 2, 3, 4, 6, 8, 9, 12, 16, 18, 24, 36, 48, 72, 144

145: 1, 5, 29, 145

146: 1, 2, 73, 146

147: 1, 3, 7, 21, 49, 147

148: 1, 2, 4, 37, 74, 148

149: 1, 149

150: 1, 2, 3, 5, 6, 10, 15, 25, 30, 50, 75, 150

151: 1, 151

152: 1, 2, 4, 8, 19, 38, 76, 152

153: 1, 3, 9, 17, 51, 153

154: 1, 2, 7, 11, 14, 22, 77, 154

155: 1, 5, 31, 155

156: 1, 2, 3, 4, 6, 12, 13, 26, 39, 52, 78, 156

157: 1, 157

158: 1, 2, 79, 158

159: 1, 3, 53, 159

160: 1, 2, 4, 5, 8, 10, 16, 20, 32, 40, 80, 160

161: 1, 7, 23, 161

162: 1, 2, 3, 6, 9, 18, 27, 54, 81, 162

163: 1, 163

164: 1, 2, 4, 41, 82, 164

165: 1, 3, 5, 11, 15, 33, 55, 165

166: 1, 2, 83, 166

167: 1, 167

168: 1, 2, 3, 4, 6, 7, 8, 12, 14, 21, 24, 28, 42, 56, 84, 168

169: 1, 13, 169

170: 1, 2, 5, 10, 17, 34, 85, 170

171: 1, 3, 9, 19, 57, 171

172: 1, 2, 4, 43, 86, 172

173: 1, 173

174: 1, 2, 3, 6, 29, 58, 87, 174

175: 1, 5, 7, 25, 35, 175

176: 1, 2, 4, 8, 11, 16, 22, 44, 88, 176

177: 1, 3, 59, 177

178: 1, 2, 89, 178

179: 1, 179

180: 1, 2, 3, 4, 5, 6, 9, 10, 12, 15, 18, 20, 30, 36, 45, 60, 90, 180

181: 1, 181

182: 1, 2, 7, 13, 14, 26, 91, 182

183: 1, 3, 61, 183

184: 1, 2, 4, 8, 23, 46, 92, 184

185: 1, 5, 37, 185

186: 1, 2, 3, 6, 31, 62, 93, 186

187: 1, 11, 17, 187

188: 1, 2, 4, 47, 94, 188

189: 1, 3, 7, 9, 21, 27, 63, 189

190: 1, 2, 5, 10, 19, 38, 95, 190

191: 1, 191

192: 1, 2, 3, 4, 6, 8, 12, 16, 24, 32, 48, 64, 96, 192

193: 1, 193

194: 1, 2, 97, 194

195: 1, 3, 5, 13, 15, 39, 65, 195

196: 1, 2, 4, 7, 14, 28, 49, 98, 196

197: 1, 197

198: 1, 2, 3, 6, 9, 11, 18, 22, 33, 66, 99, 198

199: 1, 199

MATRIX INVERSION

Matrix inversion is easily done these days with a computer algebra system. Programs like Maple, Mathematica and Maxima will quickly do a symbolic matrix inversion for moderately sized matrices. In this appendix, we present the rudiments of doing a symbolic matrix inversion by hand. It is, after all, never a good idea to become too reliant on the computer. The details of doing the calculation yourself can sometimes inspire you to see connections you would not otherwise see. In what follows, we will only assume you know what the determinant of a matrix is. If you're unsure what a determinant is, you can consult almost any textbook on linear algebra.

The particular matrix inverse that we want to calculate is called a walk generating matrix. It is defined as follows

$$\mathbf{W}(z) = (\mathbf{I} - z\mathbf{A})^{-1} \tag{157}$$

where $\mathbf{I}$ is the identity matrix, $\mathbf{A}$ is the adjacency or transition matrix for the states of an automaton, and z is just a symbolic variable. The (i, j) element of $\mathbf{W}(z)$ is equal to the generating function for the number of walks between state i and j of the automaton. You can see this from the following expansion

$$(\mathbf{I} - z\mathbf{A})^{-1} = \mathbf{I} + z\mathbf{A} + z^2\mathbf{A}^2 + \cdots \tag{158}$$

The (i, j) element of $\mathbf{A}^k$ is equal to the number of walks of length k between state i and j of the automaton.

Unfortunately, this expansion is not very useful for calculating the inverse. We can instead express the inverse in the following form.

$$(\mathbf{I} - z\mathbf{A})^{-1} = \frac{\mathrm{adj}(\mathbf{I} - z\mathbf{A})}{\det(\mathbf{I} - z\mathbf{A})} \tag{159}$$

where $\det(\mathbf{I} - z\mathbf{A})$ is the determinant of $\mathbf{I} - z\mathbf{A}$ and $\mathrm{adj}(\mathbf{I} - z\mathbf{A})$ is called the adjugate of $\mathbf{I} - z\mathbf{A}$. It is defined as the transpose of the cofactor matrix of $\mathbf{I} - z\mathbf{A}$. Call the cofactor matrix $\mathbf{C}$, then the (i, j) element of $\mathbf{C}$ is defined as $\mathbf{C}_{i,j} = (-1)^{i+j} M_{i,j}$ where $M_{i,j}$ is equal to the determinant of the matrix that remains after deleting row i and column j of $\mathbf{I} - z\mathbf{A}$.

This is best illustrated with an example. For $\mathbf{A}$ we will use the adjacency matrix for the Fibonacci automaton shown below.

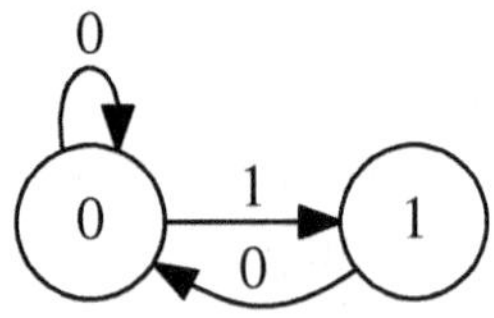

The adjacency matrix is

$$\mathbf{A} = \begin{pmatrix} 1 & 1 \\ 1 & 0 \end{pmatrix} \tag{160}$$

So we want to calculate the inverse of the matrix

$$\mathbf{I} - z\mathbf{A} = \begin{pmatrix} 1-z & -z \\ -z & 1 \end{pmatrix} \tag{161}$$

The determinant is easily calculated

$$\det(\mathbf{I} - z\mathbf{A}) = 1 - z - z^2 \tag{162}$$

To get the adjugate, we calculate $M_{1,1} = 1$, $M_{1,2} = M_{2,1} = -z$ and $M_{2,2} = 1 - z$. Putting it all together, we get the following for the inverse.

$$(\mathbf{I} - z\mathbf{A})^{-1} = \frac{1}{1 - z - z^2} \begin{pmatrix} 1 & z \\ z & 1-z \end{pmatrix}$$

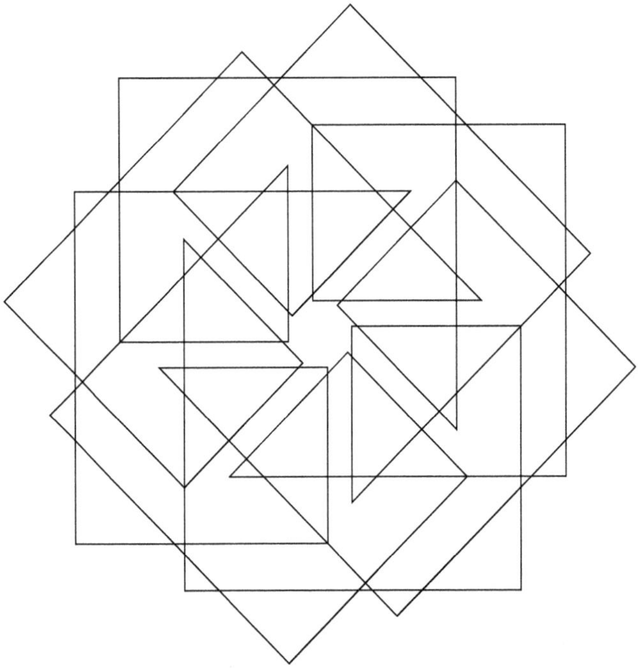

SOFTWARE

The software below can be found on this book's website:
http://www.abrazol.com/books/combinatorics2/

The programs are written in the C programming language, and will have to be compiled before you can use them. You do not have to know C to use the programs or understand the contents of the book. There is a C language compiler for every major operating system. A good one that is also free is gcc.

There is also one awk script. The version of awk that we use (gawk) runs on every major operating system.

autogen

```
Generates all words of a given length accepted
by an automaton.

Usage: autogen file.aut n s e1 e2 ...
  file.aut  = automaton file
  n = length of words
  s = start state
  ei = end state i
```

autogrep

```
Filters words accepted by an automaton.

Usage: autogrep file.aut s e1 e2 ...
  file.aut = automaton file
```

```
  s = start state
  ei = end state i
```

kbrace

```
Generates all k-ary bracelets of length n.

Usage: kbrace k n
```

kneck

```
Generates all k-ary necklaces of length n.

Usage: kneck k n
```

revfilt

```
Filters strings of length n that are unique
under reversal. Strings should be in
dictionary order.

Usage: revfilt n
```

filterones.awk

```
An awk script that filters binary words based
on a specified number of 1's.

Usage: filterones.awk numones=i
  i = number of 1's

Example:
  kneck 2 10 | filterones.awk numones=5
```

REFERENCES & FURTHER READING

- *Analytic Combinatorics*, Flajolet and Sedgewick
- *Finite Automata and Regular Expressions: Problems and Solutions*, Hollos and Hollos
- *Introduction to Combinatorial Mathematics*, C. L. Liu
- *Combinatorial Enumeration of Groups, Graphs, and Chemical Compounds*, Polya and Read
- *Notes on Introductory Combinatorics*, Polya, Tarjan and Woods
- *Generatingfunctionology*, Herbert S. Wilf
- On-Line Encyclopedia of Integer Sequences, http://oeis.org/

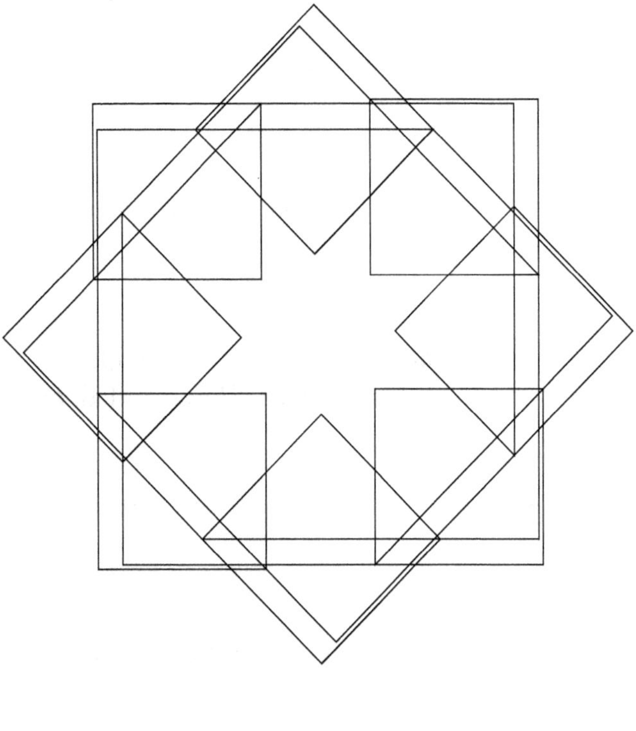

ACKNOWLEDGMENTS

In ordinary life we hardly realize that we receive a great deal more than we give, and that it is only with gratitude that life becomes rich. It is very easy to overestimate the importance of our own achievements in comparison with what we owe to others.

Dietrich Bonhoeffer, letter to parents from prison, Sept. 13, 1943

We'd like to thank our parents, Istvan and Anna Hollos, for helping us in many ways.

We thank Neil Sloane for creating and maintaining the On-Line Encyclopedia of Integer Sequences, as well as everyone else that helps keep that useful website updated.

We thank the makers and maintainers of all the software we've used in the production of this book, including: the Emacs text editor, the LaTeX typesetting system, Inkscape, mupdf and evince document viewers, Maxima computer algebra system, gcc, awk, bash shell, and the GNU/Linux operating system.

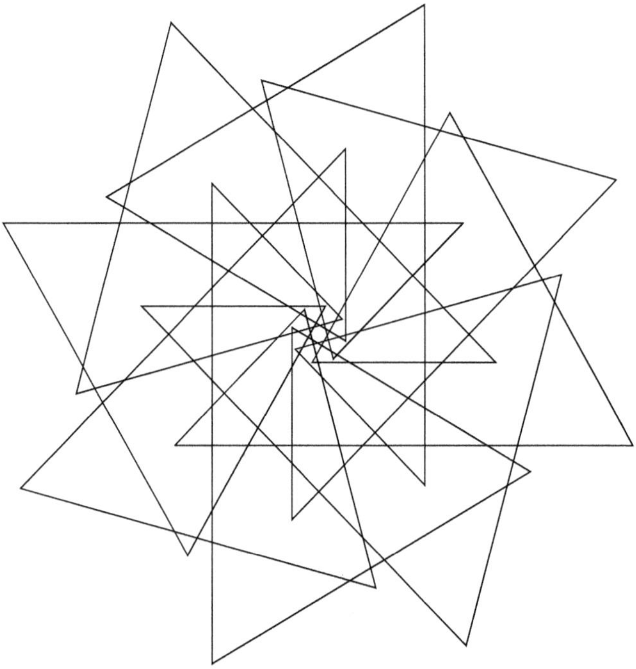

ABOUT THE AUTHORS

Stefan Hollos and **J. Richard Hollos** are physicists by training, and enjoy anything related to math, physics, and computing. They are the authors of

- **Information Theory: A Concise Introduction**
- **Recursive Digital Filters: A Concise Guide**
- **Art of Pi**
- **Creating Noise**
- **Art of the Golden Ratio**
- **Creating Rhythms**
- **Pattern Generation for Computational Art**
- **Finite Automata and Regular Expressions: Problems and Solutions**
- **Probability Problems and Solutions**
- **Combinatorics Problems and Solutions**
- **The Coin Toss: Probabilities and Patterns**
- **Pairs Trading: A Bayesian Example**

- **Simple Trading Strategies That Work**
- **Bet Smart: The Kelly System for Gambling and Investing**
- **Signals from the Subatomic World: How to Build a Proton Precession Magnetometer**

They are brothers and business partners at Exstrom Laboratories LLC in Longmont, Colorado. Their website is exstrom.com

THANK YOU

Index

www.ingramcontent.com/pod-product-compliance
Lightning Source LLC
Chambersburg PA
CBHW070514200326
41519CB00013B/2801
9781887187329